T0231725

Peer-to-Peer Computing

Applications, Architecture, Protocols, and Challenges

Chapman & Hall/CRC
Computational Science Series

SERIES EDITOR

Horst Simon
Deputy Director
Lawrence Berkeley National Laboratory
Berkeley, California, U.S.A.

AIMS AND SCOPE

This series aims to capture new developments and applications in the field of computational science through the publication of a broad range of textbooks, reference works, and handbooks. Books in this series will provide introductory as well as advanced material on mathematical, statistical, and computational methods and techniques, and will present researchers with the latest theories and experimentation. The scope of the series includes, but is not limited to, titles in the areas of scientific computing, parallel and distributed computing, high performance computing, grid computing, cluster computing, heterogeneous computing, quantum computing, and their applications in scientific disciplines such as astrophysics, aeronautics, biology, chemistry, climate modeling, combustion, cosmology, earthquake prediction, imaging, materials, neuroscience, oil exploration, and weather forecasting.

PUBLISHED TITLES

PETASCALE COMPUTING: ALGORITHMS AND APPLICATIONS
Edited by David A. Bader

PROCESS ALGEBRA FOR PARALLEL AND DISTRIBUTED PROCESSING
Edited by Michael Alexander and William Gardner

GRID COMPUTING: TECHNIQUES AND APPLICATIONS
Barry Wilkinson

INTRODUCTION TO CONCURRENCY IN PROGRAMMING LANGUAGES
Matthew J. Sottile, Timothy G. Mattson, and Craig E Rasmussen

INTRODUCTION TO SCHEDULING
Yves Robert and Frédéric Vivien

SCIENTIFIC DATA MANAGEMENT: CHALLENGES, TECHNOLOGY, AND DEPLOYMENT
Edited by Arie Shoshani and Doron Rotem

INTRODUCTION TO THE SIMULATION OF DYNAMICS USING SIMULINK®
Michael A. Gray

INTRODUCTION TO HIGH PERFORMANCE COMPUTING FOR SCIENTISTS
AND ENGINEERS, **Georg Hager and Gerhard Wellein**

PERFORMANCE TUNING OF SCIENTIFIC APPLICATIONS, **Edited by David Bailey,
Robert Lucas, and Samuel Williams**

HIGH PERFORMANCE COMPUTING: PROGRAMMING AND APPLICATIONS
John Levesque with Gene Wagenbreth

PEER-TO-PEER COMPUTING: APPLICATIONS, ARCHITECTURE, PROTOCOLS, AND CHALLENGES
Yu-Kwong Ricky Kwok

FUNDAMENTALS OF MULTICORE SOFTWARE DEVELOPMENT
Victor Pankratius, Ali-Reza Adl-Tabatabai, and Walter Tichy

Peer-to-Peer Computing

Applications, Architecture, Protocols, and Challenges

Yu-Kwong Ricky Kwok

CRC Press
Taylor & Francis Group
Boca Raton London New York

CRC Press is an imprint of the
Taylor & Francis Group, an **informa** business

A CHAPMAN & HALL BOOK

CRC Press
Taylor & Francis Group
6000 Broken Sound Parkway NW, Suite 300
Boca Raton, FL 33487-2742

Printed in the United States of America on acid-free paper
Version Date: 20110627

International Standard Book Number: 978-1-4398-0934-1 (Hardback)

Visit the Taylor & Francis Web site at
http://www.taylorandfrancis.com

and the CRC Press Web site at
http://www.crcpress.com

Contents

List of Figures

List of Tables

Preface

Peer-to-peer computing, at least on a conceptual level, is a genuine paradigm shift—intelligence is at the edge, computing is completely decentralized, and the network is just there to knit the distributed intelligence together. Indeed, with advancements in hardware technology, proliferation of the open source development culture, and abundant information at our fingertips, computing power and user competence at the edge of the network has risen to an unprecedented level. Thus, devices at the edge (not restricted to desktop PCs) can congregate and share their resources (computing power, file data, etc.) to provide services to participating users in a self-sufficient manner, *without* the need of dedicated servers. With potentially up to millions of machines participating simultaneously (e.g., when some hot events are occurring), the aggregated computing resources can dwarf any powerful server farm.

Well, well, well, ... this is "conceptual level" thinking as of now. There are still many road blocks to such a vision, even though we do see millions of machines working together in a P2P manner (e.g., streaming live video events). Again, as the old saying goes, the devils are in the details. Thinking up such a gigantic scale of shared computing resources is one thing, while implementing the idea is definitely another. Road blocks to the grand vision of truly global P2P sharing include architectural maintenance problems arising from the sheer scale of the system, incentives for truthful cooperation, trust among peers when they need to accept data from remote sources, security issues caused by the inevitable existence of malicious users, etc.

The purpose of this book is to serve as a first-rate guide to these road blocks of a grand vision. The problems arising from these different aspects of P2P computing are first described in detail in the respective chapters. This is followed by a detailed survey of proposed solutions. A major conclusion of this book is that there is still much work to do. Perhaps this should not be a surprise because in a sense we are trying to build a self-governing crowd of computers sharing resources at will. Sounds like a cyberspace version of utopia?

The target audience of this book are senior level undergraduate students and graduating students interested in P2P research. Each chapter of this small book can be read independently. Contents of the book can easily be used as materials for a one-semester course.

Many people have helped in making the publication of this book a reality, possibly unknowingly. Among them I wish to thank my two previous Ph.D.

students, Dr. Tyrone Kwok and Dr. Carson Hung, who helped in providing advice and useful information on various topics, even when they were extremely busy cranking out interesting iPhone apps in their company. Many colleagues in Colorado State University and the University of Hong Kong also provided valuable experience and information that I used in the book in numerous sections. I am also extremely grateful to Ms. Randi Cohen for her patience. Finally, I would also like to thank my wife, Fion, and children, Harold and Amber, for their tolerance of my "mental absence" (despite physical presence) and sporadic outbursts of frustration when I got stuck in the writing of the book.

Chapter 1

Introduction

1.1 Overview

Modern computing technologies have decentralized data processing power in an unprecedented manner. An important implication is that user machines, be it a desktop computer or a handheld PDA (personal digital assistant), have data processing power, in terms of instruction processing rate, amount of storage, and reliability, that was inconceivable merely a decade or two ago. Indeed, computing now occurs largely at the "edge" of networks. Network infrastructure systems have also made tremendous strides thanks to the ever improving communications technologies. Advancements in computing and communication, coupled together, enable a recent trend in a new form of distributed processing—peer-to-peer (P2P) computing [Oram, 2001, Steinmetz and Wehrle, 2005, Leuf, 2002, Minar and Hedlund, 2001, Milojicic et al., 2002, Roussopoulos et al., 2004, Schoder and Fischbach, 2003, Smith et al., 2003].

As its name implies, P2P computing involves users (or their machines) on equal footing—there is no designated server or client, at least in a persistent sense. Every participating user can be a server and be a client depending on context. Some people have referred to this as a "democratic computing environment" [Androutsellis-Theotokis and Spinellis, 2004] because users are free from centralized authorities' control. This new paradigm of distributed computing has spurred many high profile applications, most notably in file sharing, such as: BitTorrent [Cohen, 2003], Freenet [Clarke et al., 2000], Gnutella [Gnutella Protocol Development, 2009], and, of course, Napster [Napster, 2009].

Apart from the most commonly known wired P2P file sharing applications, some other wireless P2P applications have become part of our daily life. For example, people can now play numerous P2P Java online games which are compatible with mobile phones so that players are allowed to interconnect in local area through Bluetooth and WiFi, or wide area through 3G network. Indeed, in many metropolitan cities such as Hong Kong and Tokyo, we can see that train commuters routinely play wireless games among each other using popular devices such as PSPs (Play Station Portables). Now, many mobile phones also already have toward GB or higher RS-MMC card storage capability. Indeed, it is now a common practice to have P2P file transfer through "BlackBerry email service" on mobile phones. As such, wireless P2P file sharing is not only feasible but also becoming pervasive.

The highly flexible features of P2P computing such as a dynamic population (users come and go asynchronously at will, at a dramatic scale, called *flash crowd*), dynamic topologies (it is impractical, if not impossible, to enforce a fixed communication structure), and anonymity, come at a significant cost-autonomy which, by its very nature, is not always in harmony with tight cooperation. Consequently, inefficient or lack of cooperation could lead to undesirable effects in P2P computing. Among them the most critical one is "free-riding" behavior. Loosely speaking, free-riding occurs when some users do not follow the presumed altruistic cooperation rules such as sharing files voluntarily, sharing bandwidth voluntarily, or sharing energy voluntarily, so as to benefit the whole community.

Such altruistic sharing actions, presumably, would bring indirect and intangible (and even remote) returns to the users. For instance, if everyone shares files voluntarily, every user would eventually benefit from the high availability of a large and diverse set of selections. Unfortunately, there are some users that do not believe or buy in to such utopia-like concepts and would, then, "rationally" choose to just enjoy the benefits derived from the community, but not contribute their own resources. Thus, a successful P2P system requires an effective incentive providing mechanism, which is currently a very hot topic of P2P research.

Apart from incentives, there are three other major research problems faced by a P2P system. Firstly, even if a participating peer has all the incentives to cooperate, there is a trust issue that needs to be handled. Specifically, if there is no trust management system incorporated in the P2P system, it is difficult for a cooperative peer to determine whether another remote peer is trustworthy or not. For example, in a file sharing application, it can be difficult for a cooperative peer to accept a file sent from a remote peer that may not be trustworthy.

Secondly, as a P2P system scales up, performance quickly becomes an issue. Indeed, many popular file sharing P2P systems can have hundreds of thousands of users participating at the same time. The response time perceived by each peer is therefore critically determined by how efficient the P2P network can deliver the requests and results. One major factor is the network topology, which governs how the participating peers are connected among each other. Specifically, P2P networks can have a structured topology, an unstructured topology, or a hybrid between the two. Nevertheless, for all P2P systems, topology control is always needed to dynamically adjust the connectivity among peers in order to optimize the performance of the P2P applications.

Thirdly, and perhaps most importantly, there is a security issue in practical use of P2P systems. Indeed, by nature of a P2P system, peers interact without the intervention of a central authority. Thus, even if incentive and trust are successfully tackled, security issues such as confidentiality and data integrity are still notoriously hard to solve. This is because without a central authority such as a certification authority (CA), keys distribution among peers is very difficult to handle. Consequently, it is difficult to realize communication confidentiality. On the other hand, peers' communications and topology control rely very much on reliable updates among peers to maintain a consistent routing table. Yet, again without the help of a central authority, such update messages' integrity can be easily compromised by some malicious peers.

In summary, P2P systems present a unique combination of challenges listed below.

Highly Decentralized Organization. The advent of P2P systems is due to the ever increasing desire of moving away from centralized control, in both aspects of accessing computing resources and accessing information. Thus, it is very difficult, if not impossible, to coordinate the peers in an organized manner. This in turn leads to an inevitable detachment of data from the sources. Essentially, when a peer wants to access a data item or some service, it cannot target a particular "server" but a swarm of potential suppliers. As such, redundancy is intrinsic in a P2P system.

Absolute Autonomy. Every peer is autonomous and its behaviors are not under any centralized controller. A peer does not even need to follow any "protocol" but is instead "enticed" with some incentive schemes to cooperate. Indeed, a Byzantine behavior model should be assumed for an arbitrary peer. Consequently, it is difficult to deduce system performance from a bottom-up perspective. Instead, it can only be deduced from a holistic emergent angle. Finally, autonomy of peers also implies possibly malicious actions can be carried out by an arbitrary peer, exacerbating security concerns.

Possibly Unstructured Networking. From a networking point of view, although structured network topologies have been widely considered (detailed in Chapter 3), currently a mesh or random swarm networking is the norm. This is because maintaining a structured topology goes directly against the autonomy of peers. Thus, such a structured network architecture is only realized at a system level, e.g., connecting the trackers in a BitTorrent network, but not at the user level. As a result, it is difficult to provide quality-of-service (QoS) guarantees to users.

Unreliable Communication Environment. Application level networking is used among peers, and thus, the connections can be unreliable. For instance, peer dynamics (i.e., peers joining and leaving) can lead to some broken connections. Thus, similar to the lack of structured topology, it is difficult to provide QoS guarantees. Similarly, in a P2P wireless network (e.g., a wireless sensor network), the communication links among peers are also highly unreliable. Consequently, a more fault-tolerant communication paradigm has to be devised for such a P2P network.

Large Population. A P2P system usually scales to a large number of users (e.g., up to several millions of simultaneous users) and thus, data and/or peer search has to be able to handle a large user population. Indeed, when more users join, more resources are aggregated, and hence, can support even more users. This is a very unique self-scaling effect of a P2P system. As a result, some kind of hierarchy has to be used in order to cope with the scalability issue. For instance, tracker servers are commonly used for keeping track of data and peer locations.

The purpose of the book is two-fold: (1) to introduce the existing applications and technologies employed; and (2) to motivate further research issues involved.

1.2 Road Map

In Chapter 2, to set the stage for understanding the various important research issues in P2P systems, we first introduce the various P2P network architectures. In Chapter 3, we discuss the topology control research problem in detail. In Chapter 4, we provide a detailed survey on the existing technologies for handling the topology control issues. In Chapter 5, we describe various novel and interesting incentive schemes for enticing peers to cooperate. In Chapter 6, we describe the recent innovations on trust issues. In Chapter 7, we focus on the security problems in a P2P network. We provide some concluding remarks in the final chapter.

Throughout Chapters 2 to 7, we use the highly popular P2P IPTV application PPLive [PPLive, 2009, Vu et al., 2010] as a case study to illustrate the practical aspects of the concepts covered.

Chapter 2

P2P Applications

2.1 Introduction

As in many computing technology breakthroughs, the advancements in peer-to-peer (P2P) systems are largely brought about by applications' demands. Indeed, evolving from the early file sharing systems to nowadays' video streaming systems, many novel efficient solutions have been proposed and implemented to satisfy various users' requirements. Thus, before we look at these advancements in detail in later chapters, it is useful for us to review the evolution of P2P applications in this chapter.

As shown in Figure 2.1, there are three main components in a typical P2P system. The first one is a Web portal of the application, also known as a login server, which is the point of getting access to the P2P service. Essentially, the user first connects to this server in order to check out the availability of services as well as peers. The second main component is usually referred to as the tracker nowadays, which is essentially a directory server furnishing peers availability information to a new peer. In the early days of P2P computing, these two servers were typically implemented in a single system. Yet as user population grows and the service becomes much more diversified (e.g., there are numerous video channels available), the peers tracking system function has to be overloaded to a separate server, namely the tracker. The third component is of course the peers, which are autonomous client machines that join and leave the system at will. The peers serve many useful or even critical functions. For instance, one of the most important functions is that peers help each other to get the necessary data packets.

Specifically, there are three key aspects governing the behaviors of a P2P application.

Discovery. Upon entering the P2P system, the very first tasks a new peer needs to carry out are the discoveries of services, data, and peers. The peer first has to find out if a particular desired service (e.g., a certain video channel or a specific file) is available. Then the peer needs to determine the various pieces of meta-data about the service, such as the location information and the size of the actual data. Finally, before the actual data can be downloaded, the peer needs to obtain a list of peers, from the corresponding tracker, for making data transfer connections.

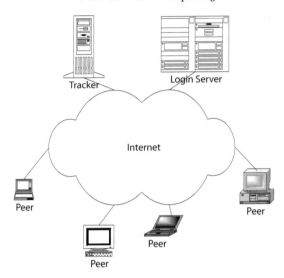

FIGURE 2.1: A general architecture of a P2P application.

Location. Here, location is used in two different senses. First, a new peer needs to obtain location information about the corresponding tracker (e.g., the tracker's IP address), about the peers that own the needed actual data. Second, the new peer also needs to report to the P2P servers about its own location and the data it already possesses. Such location information exchange is crucial for the tracker servers to keep accurate data about the availability of actual data and peers.

Data Transfer. The uploading and downloading of the actual desired data are obviously the ultimate important steps. These steps are also the aspects where different P2P systems take on different approaches. For one thing, there are the so-called *push* and *pull* approaches for data exchange. In a push-based approach, it is the data uploading peer who determines the recipients of the data. In contrast, in a pull-based approach, it is the data downloading peer who sends out transfer requests to a set of potential data senders. Another important dimension about data transfer is the network topology issue. Again there are two general approaches. The first one is a structured approach, in which the connections among peers are governed by a well-defined network topology such as using a distributed hash table (DHT). The second approach is a so-called *mesh* approach in which connections among peers are totally ad hoc and do not follow any structured topology.

More about these different key aspects of a P2P application are further explained when we describe the specific applications below.

Let us also examine the issue of application performance. As in any computing system, we can evaluate a P2P application's performance from a user-oriented perspective and a system-oriented perspective. Such a distinction between different performance metrics is even more acute in a P2P system because of the autonomous, and sometimes even "selfish," nature of the participating clients. Indeed, while each peer tries to optimize its own performance in terms of a certain user-oriented metric (e.g., downloading time), the performance of the whole system (or the whole P2P community) may be degraded by such local optimizations.

Availability. The availability metric measures the ease of getting access to the item in need, which may be data or a particular peer. To capture the concept of ease, usually availability is defined as a probability. Specifically, it is defined as the probability that the item in need can be obtained. For instance, in a P2P file-sharing system, we can compute the probability of successfully downloading a certain file in need to indicate the level availability in the system. One point we have to note is that availability is by and large a system-oriented metric because it is usually not in the interest of a peer to maximize the availability of contents in the system. To enhance availability, however, it requires participating peers' efforts to ensure that the data in need are replicated widely.

Download Time. The time it takes for a peer to successfully download a file (or a stored video) is obviously a key performance metric from a user's point of view. Download time is affected by many factors including data replication level, peer connectivity, uploading/downloading data rates, etc. To put the complexity into perspective, even if the whole system is under centralized control, it is still extremely difficult to come up with a data storage plan and peer connection topology to optimize each peer's download time.

Robustness. Peer dynamics (i.e., peers come and go) is a fact of life in a P2P system. Thus, we cannot expect a P2P system to be "stabilized" in a traditional sense. We should, however, try to design and implement the system such that it is robust to changes. Thus, similar to availability, we can define robustness as probability that the P2P system can still provide a certain level of performance (e.g., in terms of average download time, or in terms of availability) subject to a particular model of peer dynamics. Similar to availability, robustness is also a system-oriented metric, which requires peers' cooperation (perhaps unknowingly) to optimize.

Scalability. Nowadays, any Internet service has to be able to support a large user population—on the order of at least hundreds of thousands of simultaneous users—in order to be "notable." Thus, it is mandatory for a P2P system to exhibit a high degree of scalability, which can be quantified as a certain rate of population growth that can be supported while maintaining a more or less stable level of performance (again in terms of average download time, or in terms of availability). Scalability is therefore also a system-oriented performance metric.

Server Cost. As will be evident from the surveys below, a practical P2P system still needs a good number of servers to support many useful system functions, e.g., tracking existing peers, organizing the connections among peers, etc. Thus, a useful indicator about the difficulty in implementing a P2P system is the total cost required to install these servers. Another variable in the cost function is the bandwidth fees required for these servers.

In the remainder of this chapter, we first briefly survey P2P applications that are designed for computation sharing. This is followed by discussions on P2P applications for the classical file sharing service. We then move on to survey P2P media applications—voice and video. A summary is presented at the end.

2.2   Distributed Processing

2.2.1   Internet Computing

One early application of a P2P computing model is to share the processing load among many decentralized machines. The rationale is that for many "pleasantly parallel" computing problems (i.e., large scale data parallel problem with very little or even no dependency among parallel tasks), the aggregate processing power of a large number of machines can match the processing power of an expensive supercomputer. This is similar in spirit to Grid Computing [Butler et al., 2000, Foster and Kesselman, 1999].

SETI@Home [Anderson et al., 2002, SETI@Home, 2009], launched in 1999, is as yet the largest effort in distributed processing in terms of participants. It is reported that it currently has over 5 million users worldwide. The objective of SETI (Search for Extra-terrestrial Intelligence) is to exploit the aggregate computing power of a large number of computers actively linked to the Internet in order to process the daunting quantity of radio signal data gathered at the Arecibo Observatory in Puerto Rico. The "search" is in fact a detection problem in the sense that a large quantity of radio signal data are mined so as to check whether there are some unusual signals which could possibly be sent by extra-terrestrial creatures from a distant planet. A participating computer obtains signal data from the central server at Berkeley and then processes the data using a client program downloaded from the server site. Results are then sent back to the server for further analysis. As a "pleasantly parallel" computing problem, more participants in SETI@Home generally imply faster and/or more accurate results.

We can see that SETI@Home's computing model is mainly a client-server one, and thus, should probably not be considered as a P2P application. However, there is a "competitive" feature in the project in that different users can compete to earn credits by trying to produce results faster than others. Users can also form teams in order to process the signal data in a collaborative manner. Thus, we can see that SETI@Home does possess a P2P component. Furthermore, as in any competition, some users try to cheat by sending in results which are in fact not yet completely processed, by modifying the code of the client programs. Consequently, SETI@Home server also has to police the users' behaviors. These issues are commonly seen in a typical P2P application. There are still many open problems in these regulatory aspects of the project.

There are many other similar projects on large scale distributed computing [Einstein@Home, 2009, Folding@Home, 2009, BOINC, 2009]. In particular, the BOINC [BOINC, 2009] (Berkeley Open Infrastructure for Network Computing) environment is a free programming tool for users to develop other large scale distributed processing applications. One interesting feature of the BOINC platform is that it has a credit system for developers to implement a policing service for authenticating results from participating computers. As to other programming tools for implementing such large scale distributed processing, recently there is a widely considered software called GreenTea [GreenTea Technologies Inc., 2009], which is a purely Java-based P2P platform. Specifically, installed with custom-made GreenTea client programs, participating computers can share in and out their resources such as computing cycles, storage spaces, or services.

2.2.2 Wireless Sensor Networks

Wireless Sensor Networks (WSNs) [Akyildiz et al., 2002] have gained remarkable attention as they have become highly attractive distributed processing platforms, thanks to the recent advancements of electronic and wireless technologies. Due to the inherent decentralized and autonomous behaviors of sensors, a WSN can also be considered as a P2P platform. Specifically, a WSN usually consists of ultra small autonomous devices called sensor nodes, which are battery powered, limited in memory storage and computational power. In a typical application scenario, sensors cooperatively monitor physical and environmental conditions and then transmit collected data to a sink node or base station via wireless links for further analysis. For instance, in a military scenario, WSNs are deployed in a large scale with well over 10,000 nodes [Chan et al., 2003, Du et al., 2004, Eschenauer and Gligor, 2002] for gathering a large volume of target recognition data. Smart Dust [Smart Dust Project, 2008], WINS [WINS Project, 2008], and μAmps [μAmps Project, 2008] are well-known examples of WSN research projects.

The development of WSNs was originally motivated by the military sensing and tracking arena, such as battlefield surveillance. When sensor nodes are deployed in hostile areas, security becomes extremely important as nodes are subjected to different kind of threats [Fu et al., 2005, Newsome et al., 2004, Parno et al., 2005, Wang and Bhargava, 2004, Wood and Stankovic, 2002]. Nodes may be captured and the communications among them may be eavesdropped or altered. Therefore, messages transmitted between sensor nodes must be encrypted using various cryptographic protection schemes to guard against different types of malicious attacks. Hence, *trust establishment* is one of the most critical components to set up a secure communication environment in a WSN-based information system [Kwok, 2007].

Although many traditional trust establishment schemes have been proposed, a WSN is unique due to its distributed and resource-constrained properties. Some methods such as using a master key or pairwise private sharing of keys are proposed [Kwok, 2007], but they are either too insecure or impractical [Chan et al., 2005a] for WSNs. Currently, many trust establishment schemes have been developed for wireless sensor networks [Anderson et al., 2004, Kwok, 2007]. Among them, key pre-distribution schemes are widely considered as practicable solutions in WSNs [Chan et al., 2003, Chan and Perrig, 2005, Du et al., 2005, Eschenauer and Gligor, 2002, Kwok, 2007, Liu and Ning, 2003]. A typical key pre-distribution scheme works by having keys distributed to all nodes prior to deployment. Eschenauer *et al.* [Eschenauer and Gligor, 2002] pioneered this field of research by proposing a randomized key pre-distribution scheme, which relies on probabilistic key sharing among nodes using random graph theory [Erdös and Rényi, 1960].

From a system's point of view, WSNs are often regarded as a kind of Mobile Ad-hoc Network (MANET). In order to make sensor nodes cheaper and smaller so as to facilitate large scale deployment, heavyweight and computation intensive programs are not expected to be executed on the tiny sensor devices. These limitations make the design of trust establishment schemes in WSNs highly challenging. A brief summary of the constraints is described below:

1. *Energy limitation*: Sensor devices are usually small in size and battery-powered. The limited supply of power restricts the computational and communication capabilities of sensor nodes. Indeed, an effective energy conservation scheme is vital for maximizing the lifetime of operation from a single node to the entire network [Raghunathan et al., 2006]. Nowadays, a typical sensor node can operate for a week (under full operation) to several months.

2. *Memory storage limitation*: Due to the small size, sensor nodes are equipped with limited amount of memory. Apart from storing key materials, it is still necessary to store the key management program and many other applications for operation. For instance, a popular sensor

device, namely Berkeley MICA2 Mote [Crossbox Technology, 2008], has only 128 KBytes program memory.

3. *Vulnerability to attacks*: As WSNs are usually deployed in hostile environments, nodes are exposed to physical attacks by the potential adversaries. In most practical scenarios, it is possible for the attackers to compromise a node physically and take full control of the node without being detected. This is often called "node capture attack."

4. *Lack of post-deployment knowledge*: Without deployment knowledge, such as location information, the distribution of sensor nodes may follow a randomly scattered pattern. Therefore, the network topology is hard to be determined *a priori* so that the optimization of trust establishment is very difficult.

In general, the major barrier for enhancing security in WSNs is their weak computational and communication capabilities. These inherent properties make traditional cryptographic protocols, such as Diffie-Hellman key agreement [Diffie and Hellman, 1976] and RSA signature [Rivest et al., 1978] techniques, undesirable when compared to the symmetric key approaches [Perrig et al., 2002] in terms of energy and computational requirements. These public key schemes are either too computation intensive or too large to fit in the resource-constrained sensor nodes. We discuss more on sensor network trust establishment in Chapter 6.

2.3 File Sharing

File sharing is probably the real starter of the P2P computing arena. The idea is very simple—a user wants to find a certain file (e.g., a music MP3 file) and downloads it as soon as possible. Many users share the same objective and therefore there is a large aggregate pool of files for mutual sharing. Thus, in essence, there are four aspects in a file sharing P2P application:

Search. The file sharing system has to support a convenient and accurate file search user-interface.

Peer Selection. The file sharing system has to support an efficient peer-selection mechanism so as to minimize the download time.

Connection. Peers should be able to set up more or less stable data transfer connections so that file data packets can be exchanged efficiently.

Performance. The key performance metrics are download time and availability.

Napster [Napster, 2009], launched in 1999 and shut down in 2001 by court order, was among the first P2P file sharing applications. From the perspective of attracting users, Napster was highly successful in that at its peak reportedly over 26 million participants joined the network. Napster's file sharing model is not strictly a P2P one because it relies on centralized servers to host the file lists for participants to search for desired files. Specifically, the client program on a user's machine shows the file lists obtained from the connected server, and then, the user can search and select appropriate peers for downloading. Thus, the selection of peers is done manually by the user. As such, there was essentially no control over the authenticity of files and the performance of the downloading. With centralized servers at its core, the Napster network was not highly scalable. Currently, WinMX [WinMX World, 2009] is the most prominent client program that is based on the Napster related protocols.

Gnutella [Gnutella Protocol Development, 2009] represented a major improvement over the Napster model. Indeed, the Gnutella protocol is fully decentralized in that participating peers help each other in file discovery, control messages routing, and file transmission. As such, "Gnutella" nowadays hardly refers to any particular piece of P2P application. Rather it refers to a whole family of file sharing applications that are implemented based on the open Gnutella protocol. As in any fully decentralized file sharing system, when a client starts, the very first problem is to discover and locate other active peers. In the original design of the Gnutella protocol, this was based on a flooding mechanism—the starting client broadcasts the so-called "ping" messages over the network. When such a ping message is received by an active Gnutella user, it replies a "pong" message to the starting client. Obviously, a more fundamental question is that to whom should the starting client send the requests in the first place? Many heuristics are used in this bootstrapping process. For example, the starting client can use the list of well-known users that come with the client program. Another scheme is to use a Web cache of actively connected machines.

From a scalability point of view, a drawback of the blind flooding approach is that the volume of traffic generated could be large, even if the maximum hop-count a request message can travel is usually limited to 7. Thus, the notion of "ultra-peer" is introduced in the Gnutella protocol. Specifically, some participating peers are designated as ultra-peers which play the role of "hubs" or "routers" in the Gnutella network. When a new client starts, it actually connects to several (e.g., three) such ultra-peers, each of which could be connected to more than 30 other ultra-peers. Essentially, a user (as a leaf node) sends a request message to its ultra-peers which then forward to its connected ultra-peers. Consequently, with such a more hierarchical network structure, the scope that can be reached by a request message becomes much larger yet the traffic volume generated is limited.

As in Napster, when a Gnutella request finds a suitable provider for the desired file, the two peers can then manage the transfer without any other intervention. Foxy [Foxy, 2009] is one of the most popular implementations of the Gnutella protocol and is widely used in the greater China region.

eDonkey [eDonkey, 2009] was designed to be a reliable decentralized system for sharing large files such as video files, complete sets of music albums, CD images, etc. Obviously, such a large file with size in the ranges of hundreds of mega-bytes to even giga-bytes, needs a relatively longer time to download. Thus, the system has to be reliable in at least two senses: (1) the system has to be available during the whole downloading duration; and (2) the contents of the file have to be authentic. Around these two objectives, the eDonkey system incorporates two essentially pioneering features:

- Each file is uniquely identified by the hashed (using MD5) digests of its fix-sized content chunks (around 9 kbytes each). Thus, two files are considered equivalent if all the digests match, regardless of their filenames.

- Each file can then be downloaded from several sources simultaneously because the client can get different chunks from different sources.

As to searching for files, the eDonkey system relies on servers distributed on the Internet. Specifically, each client can contact one or more such servers to locate files and report the availability of files. As such, like Napster, the eDonkey system is not necessarily a purely P2P network. Furthermore, a server-list is also needed by each client to properly join the network. The most prominent implementation of the eDonkey system is the open-source eMule [eMule, 2009], which has reportedly millions of users. iMesh [iMesh, 2009] is another highly popular implementation of the protocol for sharing MP3 music files.

KaZaA [KaZaA, 2009, Liang et al., 2005] and its variants, which are based on the FastTrack protocol [giFT-FastTrack, 2009], use some "super-nodes" in a way similar to the ultra-peers in Gnutella. On the other hand, the FastTrack protocol also employs the hashing approach used in the eDonkey system for managing large files.

BitTorrent [BitTorrent, 2009, Cohen, 2003] is by far one of the most successful P2P file sharing systems. Similar to eDonkey, in BitTorrent each shared file is divided into pieces (of size 256KB each), which are uniquely identified by their hashed digests and usually stored in multiple different peers. Thus, for any peer in need of a shared file, parallel downloading can take place in that the requesting peer can use multiple TCP connections to obtain different pieces of the file from several distinct peers. This feature is highly effective because the uploading burden is shared among multiple peers and the network can scale to a large size.

Closely related to this parallel downloading mechanism is the ingenious incentive component used in BitTorrent. Specifically, each uploading peer selects up to four requesting peers in making uploading connections. The selection priority is based on descending order of downloading rates from the requesting peers. That is, the uploading peer selects four requesting peers that have the highest downloading rates. Here, downloading rate refers to the data rate that is used by a requesting peer in sending out pieces of some other file. Thus, the rationale of this scheme is to provide incentive for each participating peer to increase the data rate used in sending out file data (i.e., uploading, or, in BitTorrent's term, *unchoking*). There are other related mechanisms (e.g., optimistic unchoking), which are described in detail in [Cohen, 2003, Qiu and Srikant, 2004]. BitComet [BitComet, 2009] is one of the most highly popular implementations of the BitTorrent protocol. Tribler [Tribler, 2009] is a popular implementation of the BitTorrent protocol for sharing video files.

2.4   Voice over IP and Instant Messaging

Telephony service is an indispensable part of our daily life. Thus, as P2P systems proliferate, people start to employ this vehicle to deliver telephony service using the voice-over-IP (VoIP) technologies.

Skype [Skype, 2009] is as yet the most successful VoIP system that is globally available, judging from its huge user population—it is reported that more than 500 million user accounts have been created and more than 50 million users are active on a daily basis. Architecturally Skype is found to be very similar to KaZaA in that the Skype network also heavily relies on the core super-node network [Baset and Schulzrinne, 2006, Caizzone et al., 2008, Kho et al., 2008]. The detailed working principles behind Skype are not known for sure because Skype is built on proprietary protocols and its traffic flows are all encrypted. Below we highlight its main protocol features based on excellent experimental studies done recently [Baset and Schulzrinne, 2006, Caizzone et al., 2008, Kho et al., 2008].

Specifically, each super-node is responsible for client discovery and location, as well as traffic relaying when the clients are behind firewalls or NATs. Indeed, being able to route traffic to/from clients behind NATs is very important because it is reported [Tang et al., 2007] that over 60% of P2P users are behind some kind of NAT. Different from a file sharing application, a VoIP session involves locating the specific callee the caller wants to reach. To this end, quite contrary to a traditional client-server-based VoIP system such as those based on SIP or ITU H.263, the Skype system relies completely on the super-nodes for storing the location information of currently on-line Skype users. Thus, the location information of the VoIP system is completely decentralized.

To start, a Skype client needs to contact the login server for authentication and obtaining peers' information (e.g., whether they are online or not). Afterward, the client needs to establish connections with one or more super-nodes so as to transmit and receive VoIP data. The client can either use a cached super-nodes list or contact the several bootstrap super-nodes (which are hardwired in the Skype client program).

Skype is robust not only in the aspect of its ability to route traffic around NATs or firewalls, but also in its lean requirements on bandwidths. Specifically, for voice traffic flows, typically the total uplink and downlink bandwidth required is only around 40 kbps.

One key aspect about the protocol and architecture of Skype is that users cannot possibly (as of this writing) refuse to be a super-node. Indeed, when the Skype client program discovers that the client's machine is powerful enough in terms of machine architecture, bandwidth available, whether it is behind a firewall or NAT, etc., the client machine could be "promoted" to be a super-node. This enforcement of super-node role could be a potential problem for the robustness of Skype as there is not enough incentive provided for a client machine to serve as a super-node.

When there are multiple parties in a VoIP session, i.e., a conferencing situation, the system needs to carry out a "mixing" operating—merging several streams of voice packets for delivery to the receivers. There are several possible approaches [Gu et al., 2008] to achieve this. For example, as shown in Figure 2.2(a), each sender (i.e., a speaker in the conferencing session) uses a separate multicast tree for sending voice packets to the receivers. Here, sender A uses its own multicast tree to deliver packets to the receivers, and sender B does the same independently. That is, the "mixing" action is accomplished by each individual receiver. While this approach is straightforward in the sense that existing multicasting infrastructure of each sender can be used independently, the obvious drawback is that there is a need to maintain a potentially large number of such trees.

Another approach is to designate one participant (or even a dedicated server) to handle the mixing and the subsequent distribution, as shown in Figure 2.2(b). This is a preferred design option in many commercial conferencing systems due to its ease of management. However, with a designated node to handle the mixing and distribution, there is obviously a limit of how many participants it can handle. Indeed, it is reported that even in Skype, the system can only allow five simultaneous participants in a conferencing session.

Toward the other extreme is the approach that uses only one single multicast tree throughout. That is, the same multicast tree is used for any sender and all the mixing and distribution of packets, as illustrated in Figure 2.2(c). Here, we have node A as the root of this unified multicast tree, and hence, when A speaks, its packets are delivered along the tree to every other participant. Now, the other speaker, node B, uses the same tree for its packets. Specifically, when B sends its voice packets, it treats itself as the root of the tree and thus, it transmits its packets to A, C, and F. At this point, the voice packets of A and B are mixed along the tree edges AD, DE, BC, and BF. More importantly, we can also see that there is a potential problem caused by asymmetric incoming and outgoing bandwidths of each node. Consider node B and we can see that it may not have sufficient outgoing bandwidth (as in a common ADSL situation) to support the mixing of packets from both A and

B for distribution to A, C, and F. Essentially, using a single tree might be a suboptimal arrangement.

Thus, it is proposed [Gu et al., 2008] that a "decoupled" approach is used, as illustrated in Figure 2.2(d). Here, we have two logical trees—one for mixing and the other for distribution. It is important to note that the constituents of these two trees are designated from the set of participants (i.e., A, B, C, D, E, and F). Indeed, in this approach, a node needs to overload itself to play potentially two to three roles (i.e., as a participant, a mixer, and a distributor). Now, when nodes A and B speak simultaneously, their packets get transmitted along the mixing tree. The final mixed outcome will then be passed to the distribution tree, from which all the participants get the packets. The key advantage of this approach is that based on the knowledge of the system capabilities (e.g., bandwidth) of the participants, two optimal trees can be determined to support the conferencing operations.

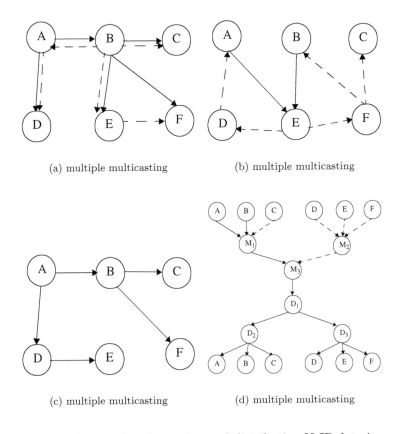

(a) multiple multicasting

(b) multiple multicasting

(c) multiple multicasting

(d) multiple multicasting

FIGURE 2.2: Approaches for mixing and distributing VoIP data in a multiparty scenario [Gu et al., 2008].

2.5   Video Streaming

Video applications are among the most important services. As such, researchers have exerted a great deal of effort in the past decade so as to achieve significant progress in providing various kinds of video services (e.g., real-time live streaming, on-demand viewing, etc.) over the Internet. Due to the considerably intensive resource requirements imposed by video streaming services, it is widely envisioned that the traditional client-server-based system does not scale (e.g., to the population size with millions of simultaneous viewers) and using a P2P approach is inevitable. Indeed, while many commercial efforts have been launched on the Internet for delivery video or TV services, the server cost is still prohibitively high, from a profit-making business's standpoint. As a case in point, it is reported [Huang et al., 2007] that YouTube [YouTube, 2009] pays in excess of 1 million dollars of bandwidth costs in providing its video-on-demand service. On the other hand, as users' machines get more and more powerful and equipped with high speed Internet connections, a P2P video data delivery model seems much more economical. Indeed, it is found [Lu et al., 2007b] that a typical ADSL user has more than 1.5 Mbps download bandwidth and over 384 kbps upload bandwidth. Such communication capabilities are good enough for supporting video services with reasonably good user experience.

The first critical component in a video streaming service is the multiple description coding (MDC) system [Goyal, 2001, Akyol et al., 2006, Wang et al., 2005] for the video contents. Simply put, with a MDC encoder, a video (i.e., a stream of picture frames) is encoded into several different layers, with different importance as to restoring the original contents at the viewer's machine. For example, consider using an MPEG-2 encoder (as in many practical systems). All the I-frames can be encoded as the first layer, which is of the highest importance. The first GOP P-frames are then encoded as the second layer. The second GOP P-frames are similarly encoded as the third layer, and so on. The B-frames can be encoded as even higher layers, which are relatively less important in restoring an acceptable video. These different layers can then be further broken down into equal size chunks (sometimes called description chunks), which are encapsulated in network packets. The most important implication of using an MDC encoder is that the more packets a node receives, the higher the quality the video playback will be.

The next critical component in a video streaming system is the engine for retrieving, storing, and playing back video packets. A generic architecture of such an engine is shown in Figure 2.3 [Hei et al., 2007b]. Here, the very first feature of such an engine is the usage of a data structure called the buffer map. First and foremost, we should note that all the chunks in the original video packet stream are labeled with unique chunk ID so that each node can keep track of any missing packets in its playback buffer. Thus, a buffer map is the data structure indicating the presence/absence of video chunks in the node. To ensure smooth and high quality playback, the playback buffer should ideally be filled with all the necessary packets. However, due to various adverse operating conditions (e.g., network congestion, etc.), some packets might be missing. Indeed, if a missing packet has not yet met its deadline (i.e., the playback time), then the node should try its best to retrieve it from somewhere. Here is where a P2P architecture could help. Specifically, in a P2P environment, a participating peer could try to request for missing packets from its connected peers. To do so in an efficient manner, the requesting peer should not blindly ask for the missing packets from all its peers. Instead, it should check whether its connected peers really possess the missing packets first before asking. Here is where the buffer maps come in handy. As can be seen in Figure 2.3, the peers periodically exchange their buffer maps so that they can tell if a connected peer really has a particular packet that is missing locally. To facilitate such P2P sharing, we can see that each node keeps two separate (logical) buffers: the playback buffer and the packet cache. While the former is used by the player for rendering the video, the latter is used for sharing with peers, and as such, might contain packets that are no longer needed locally.

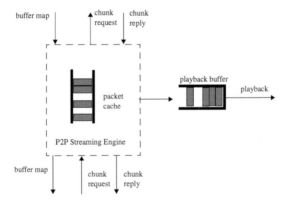

FIGURE 2.3: A generic architecture of a P2P video streaming engine [Hei et al., 2007b].

Now, obviously the next question is how a node is connected to other peers in the first place. Regarding this topology issue, there are in general two approaches: tree push and mesh pull. In a tree push approach, peers form a multicast tree so that peers owning packets needed by other connected peers can proactively send such packets to those peers along the tree [Yiu et al., 2007]. While this is efficient and can lead to a smaller delay, the tree structure itself might be too fragile given the peer dynamics (i.e., peers come and go, or called peer churn). As always, some people suggest another extreme which is the mesh pull approach. Simply put, a mesh pull approach means highly dynamic or even random connectivity. That is, each peer is connected to a dynamically changing set of peers in an unstructured way (i.e., no tree whatsoever). Consequently, a higher communication overhead would be incurred when packets are shared. Given their complementary nature, the two extreme approaches are naturally combined to form a hybrid approach [Li et al., 2008], which works by first using mesh pull to jump start the buffering process, and then construct multicast trees with relatively stable peers (or sometimes, peers with similar capabilities, as in the case of BitTorrent) to enhance the efficiency. Indeed, as pointed out by Zhang et al. [Zhang et al., 2007], using a hybrid push-pull or even a traditional mesh pull mechanism can lead to optimal performance in terms of peer upload capacity utilization and system throughput even without intelligent scheduling and bandwidth measurement.

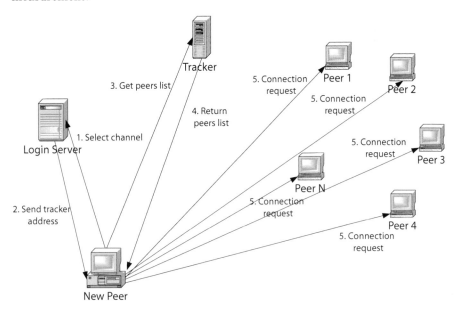

FIGURE 2.4: A general information exchange process in a P2P video streaming system.

In summary, the general process of a P2P video streaming session is illustrated in Figure 2.4. Initially, the new peer visits the so-called log-in server (i.e., the Web site of the system) to select the channel or movie the user wants to watch. The log-in server then redirects the new peer to a particular tracker server which can furnish a list of peers currently watching the same channel to the new peer. Usually the tracker server just randomly picks a subset of peers to form a list for the new peer. The new peer then selects a subset from this list so as to make connection requests. Such selection is, in current implementations, also based on randomization. After connections are established, buffer maps exchange and video packets downloading can be carried out. This general process is the basis of many well-known P2P video streaming systems such as Joost, SopCast, GridCast, UUSee, etc.

Nevertheless, there are still some differences among different systems. Indeed, Huang *et al.* [Huang et al., 2008] give a detailed analysis of design choices in a P2P video streaming system. Specifically, apart from the push versus pull architecture discussed above, the design space can be further characterized in the following several dimensions.

Chunk Size. A movie file (or a stream of live video packets) can be divided into a hierarchy of data units of different sizes. First of all, the term *chunk* refers to the largest unit, of a size around 2 MB, to be used for buffer map construction. The rationale is that the buffer maps themselves are exchanged frequently and thus, cannot be of a large size. For example, with a 200 MB movie file and a chunk size of 2 MB, each buffer map is just a 100-bit vector. The next level of storage unit is commonly called *piece* which is of a size of around 16 KB. A piece is a basic unit for the media player's processing, representing a reasonably long viewable segment of the video. However, a piece is still too large to serve as a basic unit of data requesting and downloading. Thus, in many real-life implementations, a smaller unit called *sub-piece* is also used. A sub-piece is of a size of around 1 KB, representing a moderately large size for efficient transmission (weighed against the transport protocol overheads) while not too large to cause a high loss rate.

Replication Strategies. This design aspect concerns the usage of the storage (i.e., disk space) locally in each peer. Specifically, a peer can store just one single movie at a time (referred to as *single video cache* (SVC)) or multiple movies at a time (referred to as *multiple video cache* (MVC)). Obviously, implementing a SVC is much simpler because by default the cache should store the movie currently being played back. On the other hand, implementing MVC entails at least two design considerations: (1) selection and replacement of videos; (2) pre-fetching. Firstly, we need to decide which other videos to download while the user is watching one specific movie. When the cache is full, we also need to decide which existing movie should be replaced. Secondly, we can carry out pre-fetching of other movies when the user is watching one. Yet pre-fetching other movies can also interfere with the downloading of packets for the current movie.

Chunk Selection. Given a certain status of the buffer map, a peer has to decide which missing chunks should be downloaded first. Commonly used heuristics include: sequential, rarest first, and anchor first. In the sequential selection method, the peer simply tries to get the chunks that are closest to the current playback point. In the rarest first strategy, however, the peer tries to get the currently rarest chunks in the hope that such chunks will not be "extinct" when the playback times come. Doing this will also enhance the overall availability of packets. In the anchor-based strategy, the peer tries to get all the chunks located at pre-defined anchor points used for supporting fast-forward and fast-rewind functions.

Transmission Strategies. The most important goals of transmission of video chunks are to maximize download rate and to minimize overheads. However, these goals are often times conflicting with each other. The reason is that a commonly used efficient way to maximize download rate is to request chunks from multiple neighboring peers at the same time. While such parallel downloading can speed up the transmission process, inevitably some chunks are received in duplicates, thereby leading to overheads. Indeed, in practical video streaming systems, a downloading peer usually employs one or more of the following three strategies:

- Downloading the same chunk from multiple peers at the same time;
- Downloading several different chunks from multiple peers at the same time; and
- Downloading all chunks from one selected peer and switch to another peer if the former is unavailable.

In the following we go through a brief survey of representative P2P video streaming systems.

SplitStream [Castro et al., 2003b] is one of the pioneering research-based systems for P2P video streaming. As in most early days' systems, it is designed based on a structured approach in that it uses multicast trees at the application level for distributing video packets. Specifically, to build robustness into the system, a multi-tree approach is used in that the source partitions the video into multiple stripes and then uses multiple interior-node-disjoint trees to carry each stripe to the clients. The key property of such a forest of interior-node-disjoint trees is that each participating node serves as an interior node in one and only one tree. Consequently, even if a node fails or departs abruptly, only one stripe of video is affected. Such impairment can be furthered masked by proper use of redundant coding in the MDC video.

CoolStreaming [CoolStreaming, 2009] represents one of the major pioneering efforts in large scale video streaming based on P2P technologies [Li et al., 2007, Sentinelli et al., 2007, Venot and Yan, 2007, Xie et al., 2007, Zhang et al., 2005b]. The original version was implemented in Python in 2004 and became an instant hit world-wide. Its design has also inspired many other systems, including commercial products. Basically, in CoolStreaming, a client starts by contacting a bootstrap node to obtain a list of currently active peers. It then randomly selects a subset of such peers as partners. The client can then start getting required packets from such partners using a standard mechanism, i.e., exchanging buffer maps and then getting missing packets from the partners. One distinctive feature in CoolStreaming is that once a partner, also known as a parent, receives a request from the client, it will continuously push packets to the client unless the latter decides to drop this partner. This feature is designed for enhancing the downloading efficiency.

PPStream [PPStream, 2009] is also a highly popular P2P video streaming application in the Greater China region. It is reported that with around 65 million media streaming clients every month, PPStream's share is roughly 35.1% [Wei and Chen, 2008]. PPStream's design and implementation are also rather standard. When a client starts, it first contacts a channel list server node (cached or from a list of well-known bootstrap nodes) to obtain a channel list. Upon selecting a particular channel to watch, the node then contacts a corresponding tracker node to obtain a list of peers. It then selects a subset of such peers to establish connections. Buffer maps are then retrieved from such peers and the downloading process can begin. After getting enough packets for the playback buffer, the video player can start while the node continues to share packets with other peers watching the same channel.

PPLive [PPLive, 2009, Vu et al., 2010] is one of the highly popular mesh-pull-based P2P video streaming systems. It is reported [Hei et al., 2007a] that PPLive serves on the order of 3 million daily users on average with over 300 channels at a data rate of around 250 kbps to 400 kbps. The protocols used in PPLive are observed to be similar to a typical mesh pull system. Performance-wise, PPLive still has a number of deficiencies [Hei et al., 2007a, Vu et al., 2010], despite its high popularity. For example, the start-up delay (i.e., the time duration between the time a user chooses a channel and the time the video starts) is on the order of 20 seconds for popular channels, and can go up to a couple of minutes for unpopular channels. Another problem is the high playback lags among peers. Specifically, due to different buffering progresses among peers, a peer may view the video that is lagging behind some other peers. It is found [Hei et al., 2007a] that the lag can be as high as 140 seconds of video, possibly leading to frustrating user experiences (e.g., imagine the users are watching a live soccer game). We further discuss the application properties of PPLive in Section 2.7.

Parvez *et al.* [Parvez et al., 2008] presented a useful mathematical characterization of P2P video downloading mechanisms based on a simple model. In the model, a video file is divided into M chunks, encoded at a playback rate of r. Each peer is capable of making a maximum of U connections for uploading simultaneously. A peer can also use up to D downloading connections. The average data rate of each of these connections is C. The number of downloading peers and uploading peers are denoted by x and y, respectively. New downloading peers enter the system at a rate of λ, while chunks-supplying (i.e., sources) peers stay in the system for a time period of $1/\mu$. Because each new downloading peer becomes a source itself at a rate of $(x+y)UC$, we have:

$$\frac{dx}{dt} = \lambda - (x + y)UC \qquad (2.1)$$

On the other hand, the uploading peers depart at a rate of μy and we have:

$$\frac{dy}{dt} = (x + y)UC - \mu y \tag{2.2}$$

Solving the above equations for an equilibrium solution, i.e., $\frac{dx}{dt} = \frac{dy}{dt} = 0$, we have:

$$x = \lambda(\frac{1}{UC} - \frac{1}{\mu}) \tag{2.3}$$

and

$$y = \frac{\lambda}{\mu} \tag{2.4}$$

The above results have the following implications.

- The population size of downloading peers and source peers is linearly dependent on the peer arrival rate.

- The population size of source peers is directly proportional to the residence time ($\frac{1}{\mu}$).

- The total population size is independent of the source residence time.

Most importantly, the downloading latency T can be determined by using Little's Law:

$$T = \frac{x}{\lambda} = \frac{1}{UC} - \frac{1}{\mu} \tag{2.5}$$

Hei, Liu, and Ross [Hei et al., 2007b] conducted a useful study on the video quality of a P2P streaming client based on an interesting methodology. Specifically, they deployed, in a widely dispersed manner, a number of PPLive clients which are associated with packet-sniffers. Because PPLive is based on proprietary protocols, the sniffers are used for extracting information from the sniffed packets. The key information extracted is the buffer maps exchanged among the peers. Using the buffer maps, the video quality of those connected peers can be inferred by checking the continuity of the video and the delay jitter.

Feng and Li [Feng and Li, 2008] proposed a mathematically sound scheme for P2P video streaming using the network coding [Yeung, 2008] concept. Simply put, network coding works by having the sender transmit combinations of data (e.g., with the XOR operator) to the receiver. Upon receiving distinct combinations of the data, the receiver can then decode them to extract the original data. The key advantage is that the robustness of the transmissions, in a holistic manner, is greatly improved. Now, to apply the network coding idea in P2P video streaming, the following protocol is used. Specifically, if a source peer needs to transmit a chunk to a downloading peer, it further divides the chunk into m blocks. The source peer then transmits linear combinations of such blocks with random coefficients. The downloading peer, upon receiving

m linearly independent combinations of blocks, can reconstruct the original chunk using Gaussian elimination.

As we have seen earlier, currently different video streaming systems employ slightly different methods in "matching" a new peer with a subset of existing peers watching the same channel. It is, however, not difficult to envision that a probably more effective approach is to consider the "peer selection" step as an optimization problem—to minimize the download time, for example. Indeed, it would be ideal, from a holistic perspective, if such a distributed peer selection mechanism can achieve:

- Minimum download time for each peer;

- Minimum upload bandwidth required for each peer;

- Maximum content availability; and

- Maximum robustness.

Unfortunately, it is obvious that the above goals are conflicting among each other. Specifically, while the first two goals are "user-oriented," the latter two are "system-oriented." Putting this in a more incentive-related context, we can say that the first two goals are individualistic while the latter two concern social welfare.

It is highly likely that, therefore, any peer selection scheme would not be able to provide optimal results for all of the above goals. Thus, a more practical research question is whether we can come up with some efficient distributed heuristics to achieve near-optimal results for all or some of the above goals.

Bonald *et al.* [Bonald et al., 2008] consider a "push-based" peer selection problem. Specifically, in a push-based approach, it is the chunk sender who selects requesting peers as receivers of its chunks. Let us denote the sending peer as s and a receiving peer as r. Furthermore, let $C(s)$ and $C(r)$ denote the set of chunks currently owned by s and r, respectively. Several selection heuristics are considered:

Random Peer. A receiving peer is randomly chosen from among the list of currently connected peers.

Random Useful Peer. A receiving peer is randomly chosen from those connected peers r where $C(s)/\ C(r) \neq \phi$. Thus, this is different from the Random Peer heuristic in that a receiving peer is one which really needs some chunks from the sender.

Most Deprived Peer. A receiving peer is randomly chosen from those connected peers r where $|C(s)/\ C(r)|$ is the largest. The rationale of this heuristic is to give a higher priority to a peer that needs the most chunks from the sender.

Latest Blind Chunk. This is chunk selection heuristic. Specifically, the sender selects the video chunk with the latest time-stamp in its $C(s)$ set. The sender then selects a receiving peer using Random Useful Peer or Most Deprived Peer heuristics.

Latest Useful Chunk. The sender selects the chunk c with the latest time-stamp such that $c \notin C(r)$ for at least one of its currently connected peers r.

Random Useful Chunk. The sender randomly selects a chunk c such that $c \notin C(r)$ for at least one of its currently connected peers r.

The above basic heuristics actually define a wide design space for a push-based peer communication scheme. For instance, by combining a chunk selection scheme and a peer selection scheme, we can specify a mechanism for peer chunks transmission, e.g., Most Deprived Peer and Latest Useful Chunk. Indeed, Bonald *et al.* carried out both analytical and simulation studies to examine the efficacy of several such combinations. In their findings [Bonald et al., 2008], combinations such as Most Deprived Peer/Latest Useful Chunk and Latest Blind Chunk/Random Useful Peer are found to be highly effective in distributing the video packets to all peers in the system.

2.6 Discussion

As the above brief surveys have indicated, there is a proliferation of highly popular P2P applications catering for both discrete and continuous data. Indeed, as P2P technologies continue to advance, it is highly probable that most of our network computing services could be supported in a P2P computing model in the future. This is a trend that some people refer to as a form of "liberation" from the traditional authoritarian computing model. Some people even argue that this is inevitable because of economic forces—a centralized computing model can never keep up with the growth of user population and the associated data/computing needs.

Table 2.1 summarizes the features of several popular P2P applications.

One missing piece from Table 2.1 and our brief surveys above is the 3-D streaming service [Sung et al., 2008]. While it has become highly popular in movie theaters, 3-D video delivery on the Internet is still largely inside research labs only. There are many technological obstacles, even in a traditional client-server delivery mode. For instance, the amount of data needed for rendering a 3-D object is still daunting, not to mention an animation of such objects in real-time. Yet speaking of scale, a P2P computing model is suitable for handling large-scale data sharing. Thus, it is hopeful that a P2P 3-D streaming system is a "natural" implementation of the idea. However, another challenge

TABLE 2.1: A qualitative comparison of different P2P applications.

Name	Type	Protocol	Architecture	Additional Information
BitComet [BitComet, 2009]	File sharing	BitTorrent	Structured	Closed source; adware
eMule [eMule, 2009]	File sharing	eDonkey2000, Kad	Decentralized	GPL open source; has a large user space
GreenTea [GreenTea Technologies Inc., 2009]	Networked computing	Proprietary	Hybrid	Closed source; commercial product
KaZaA [KaZaA, 2009]	File sharing (music)	FastTrack	Hybrid	Closed source; adware/spyware
iMesh [iMesh, 2009]	File sharing (music and video)	eDonkey2000, FastTrack, Gnutella	Hybrid	Closed source; freeware; supports social networks, purchase of copyrighted materials
Joost [Joost, 2009]	Multimedia streaming	P2PTV	Hybrid	Closed source; freeware; delivers near-TV resolution images; ad-supported service
Napster [Napster, 2009]	File sharing (music)	Napster	Centralized	Closed source; freeware; acquired by Roxio in 2003, providing paid music service without P2P technology
PPLive [PPLive, 2009]	Multimedia streaming	P2PTV	Hybrid	Closed source; freeware; ad-supported service
Skype [Skype, 2009]	Voice-over-IP (VoIP)	KaZaA-alike	Hybrid	Closed source; freeware; offers paid service to initiate and receive calls via regular telephone numbers
Tribler [Tribler, 2009]	File sharing (video)	BitTorrent	Structured	GPL open source; incorporates a keyword search protocol; supports social networks for content recommendation
WinMX [WinMX World, 2009]	File sharing (music)	OpenNap, proprietary WPNP	Centralized	Closed source; freeware; official WinMX central servers were shut down in 2005

is the synchronization of data transmission and rendering. While on-demand video streaming needs to handle only one dimension—time, 3-D streaming requires handling simultaneously four dimensions—3-D space plus time.

2.7 Case Study: PPLive

PPLive [PPLive, 2009, Vu et al., 2010] is a highly popular P2P IPTV application. It is heavily used, in particular, in China. It has been reported [Vu et al., 2010] that the daily average user population is close to 1 million. Among PPLive's available channels are more than 100 Chinese TV stations, about 300 live channels, and over 20,000 video-on-demand programs (i.e., moviews). The PPLive client program is free but a closed source. As discussed earlier, the PPLive system divides the video data (live or stored) into chunks. Each channel (or movie) is shared by one distinct overlay. At the user interface level, each user can join one overlay at a time (i.e., view one channel at a time). However, at the system level, each peer machine can be participating in multiple overlays (i.e., downloading/uploading contents for channels that are not being viewed). Each PPLive client program opens a pair of TCP and UDP ports for each channel in order to communicate with the PPLive infrastructure servers (i.e., the channel management servers, the group management servers, etc.) and other peers. Currently, PPLive supports a large variety of client platforms including mobile gadgets such as Android phones, iPads, etc.

2.8 Summary

In this chapter, we first delineate the system components and performance metrics. We then walk through brief surveys of P2P applications in the areas of distributed processing, file sharing, voice-over-IP services, and video streaming. While there is a proliferation of P2P applications, there are still much more exciting developments to come because many P2P applications are still far from perfect (e.g., many server machines are still needed to support their operations) and some important applications (e.g., 3-D streaming) are still not implemented in a P2P manner.

2.9   Review Questions

1. What are the key components of a typical P2P application? Describe their functions.

2. What are the important performance metrics for a P2P application? Identify them as user-oriented and system-oriented metrics.

3. What kind of compute-intensive problems are suitable for a P2P implementation?

4. Why is a centralized directory approach (e.g., as in Napster) unsuccessful for a file-sharing application?

5. Why is it necessary for BitTorrent to include a random unchoking mechanism?

6. How does a P2P VoIP system (e.g., like Skype) allow peers to connect among each other even if they are behind firewalls or NATs?

7. Explain what are push-based and pull-based approaches in a P2P video streaming system.

8. Why is MDC (Multiple Description Coding) a key component in a P2P video streaming system?

Chapter 3

P2P Network Architectures

3.1 Introduction

It is not a severe exaggeration to say that a P2P application is all about communications at the application level, which brings about the high degree of decentralization and autonomy. Indeed, by its nature, a P2P application is about voluntary high-level sharing of resources, in terms of data possessed, storage space, and bandwidth. Yet, on the flip side, such sharing can be realized only by efficiently communicating among the peers, with a lack of infrastructure support. Communications, in turn, can be effective only if we can somehow realize a well-designed network architecture at the application level [Schollmeier, 2002].

So what do we mean by network architecture here? We use this term here to refer to how the participating peers and/or servers, at the application level (instead of at the network level), connect among each other, and how they need to carry out their obligated tasks to maintain the network. In this sense, to specify a network architecture, we do not just define its topology but also the interactions among the nodes. We can then broadly classify possible network architectures into two types: structured and unstructured. The former entails not only an articulated topology (e.g., every node has a fixed degree) but also a set of strict protocol actions for each node to carry out in order to maintain the network. Notable examples in this class are various distributed hash table (DHT) systems. The latter entails highly carefree connections—every node just autonomously decides to whom and when to connect to other peers. In a sense, there is no formal "design" per se. Many practical P2P applications, such as Gnutella and BitTorrent, are of this kind.

Ideally, an effective network architecture should possess the following nice features:

Provably good performance. It would be nice if the network communication performance is provably good. For instance, if we can derive a performance bound on the number of messages or number of hops that are needed in a certain communication scenario, then we can deduce the worst case delay of downloading a file, or locating a peer, etc.

Low network maintenance overhead. Peers come and go. Thus, the network needs efforts to maintain its "shape." If the overhead of maintaining the network's characteristics (e.g., diameter, etc.) is low, then service disruptions would be reduced to a minimum.

QoS provisioning. Combining the above two nice features, we could possibly go one step further—to provide certain guarantees of QoS to peers. For instance, it would be nice if we can guarantee a certain playback delay for every new P2P video streaming peer.

Autonomy. By its nature, each peer in the network is autonomous, in the sense that it is under the user's full control. Consequently, if the network architecture allows for such autonomy to the greatest extent, then more users would like to participate.

Robustness. The network should also be resilient to changes, e.g., peer dynamics, peer failures, network failures, etc.

Yet it is clear that the above nice features are conflicting, if not totally mutually exclusive. For example, allowing for maximum autonomy is in conflict with QoS provisioning, in most cases. Furthermore, considering a structure network such as a DHT, it is mandatory for each peer to adhere to some protocol rules (e.g., responsible for handling a pre-defined range of data items) in order to realize the nice feature of provably good performance. Thus, autonomy is sacrificed in this case. On the other hand, a totally unstructured network allows for maximum autonomy. Yet such a system can hardly provide any QoS guarantee or any provably good performance. Consequently, a "holy grail" research problem is to come up with low-overhead mechanisms to provide QoS guarantees in an unstructured autonomous P2P network architecture.

Furthermore, there are several other nice features that we must consider if we consider the existence (inevitable) of malicious users. These features are incentives, trust, and security. Incentives are important to encourage cooperative and constructive, instead of malicious, behaviors. Trust is necessary among peers in order to accept the data in exchange. Proper security measures are needed if there are really some malicious users participating in the network, possibly trying to undermine the normal operations of the system. These are the topics of later chapters in this book.

In the following we provide a brief survey of different network architecture designs.

3.2   Structured P2P Systems

In this section, we explain why structured network architecture came about and describe several representative approaches.

From a highly theoretical perspective, a structured network architecture is good in terms of network performance because it is undeniably easier to quantify and deduce the optimal structures that should be used in a certain P2P scenario. Thus, motivated by such appeal of provably good performance, early P2P systems are mainly based on structured network architectures.

As we have seen in Chapter 2, many early P2P systems, most notably video streaming systems, rely on a tree-based network architecture for peer connections and data transmissions. Presumably this is because early P2P systems were designed based on a traditional multicasting communication model, albeit at the application layer.

While a tree network architecture is efficient in terms of data replication (i.e., the wasted overheads of transmitting redundant packets), it is difficult to set up and maintain in a highly dynamic P2P application scenario. Here is where theory and practice do not work with each other too well. Indeed, while we can appreciate the theoretical merits of a tree, in practice such benefits are hard to obtain.

As a case in point, using a tree for data transmission leads to at least a couple of problems, not to mention the various maintenance issues. First, using a tree can be unfair because the leaf-nodes do not need to contribute in data sharing. Second, by the same token, the internal nodes, especially those situated high up in the tree, need to take up disproportionately large workload in terms of outgoing bandwidth. From a maintenance point of view, such internal nodes can also make the whole system vulnerable to service disruptions if they abruptly depart or crash.

Thus, taking an early video streaming application SplitStream [Castro et al., 2003b] as an example, redundancy is incorporated in the network architecture in the form of using multiple trees. Specifically, each tree is used for transmitting just a single layer of video packets, called a stripe. Different trees, forming a forest, are internal-node-disjoint so that the fairness issue is taken care of. The key property of such a forest of interior-node-disjoint trees is that each participating node serves as an interior node in one and only one tree. Consequently, even if a node fails or departs abruptly, only one stripe of video is affected.

Nevertheless, maintenance of the trees still represents a significant overhead, especially when peer dynamics is vigorous.

Before we set off to describe some unstructured approaches, we first introduce a couple of classical examples of a highly prominent structured distributed network architecture, commonly known as distributed hash table (DHT).

In the following we briefly introduce two classic DHT designs—Chord [Stoica et al., 2001a] and CAN [Ratnasamy et al., 2001]. For other notable DHT schemes (such as Pastry [Rowstron and Druschel, 2001a] and Tapestry [Zhao et al., 2004]), the reader is referred to the respective research literature.

3.2.1   Chord

Chord [Stoica et al., 2001a] is among the first in the pioneering efforts in the design and implementation of robust P2P network architecture for data storage and lookup. Chord is an ingenious improvement of the consistent hashing [Karger et al., 1997] idea, which already has at least a couple of nice features. First, with high probability the hashing action provides a balanced load over the network in the sense that each node receives more or less the same amount of data items for storage. Second, with high probability when the Nth node joins (or leaves), only an $O(1/N)$ fraction of data items need to be moved to a different node in the network. However, it originally has a requirement that every node has to know about any other node in the network. If this requirement is removed, to look up an item, potentially (in the worst case) all N nodes in the system have to be checked. This is obviously a big obstacle for a higher scalability. Chord improves this by requiring each node to store only $O(\log N)$ information about other nodes, and consequently, a lookup requires only $O(\log N)$ messages. A join or leave event also entails only about $O(\log^2 N)$ messages.

In consistent hashing, a hash function, e.g., SHA-1, is used for hashing an input numeric to an m-bit identifier. The input numeric can be the IP address of a node, the contents of a file, etc. The only requirement about the parameter m is that it must be large enough so that the hashed outputs have negligible probability of being equal. Now, as everything is turned into an m-bit identifier, we can consider an identifier space organized as a ring—a circular list of numbers from 0 to $2^m - 1$ (so that we have to use modulo arithmetic with base m). For any data item (e.g., a file) with an identifier k (i.e., its hashed output is k), it is mapped to a node (i.e., a participating machine) with identifier equal to k or the first node with identifier following k (notice that in practice, definitely not every identifier in the ring will have a machine participating currently). Such a node is then defined as successor(k).

Figure 3.1 shows an example with $m = 3$ [Stoica et al., 2001a] and currently there are three nodes participating (i.e., nodes 0, 1, and 3). Now, if we need to store three data items with identifiers 1, 2, and 6, then we need to store data item 1 to node 1, data item 2 to node 3, and data item 6 to node 0, according to the above definition of successor relationship.

As a minimum requirement, each node only needs to know (i.e., keep a piece of state information such as IP address, etc.) its successor node in the ring in order to accomplish the process of data lookup. However, as mentioned above, in the worst case a lookup may traverse all the N nodes currently in the network. Here is how Chord ingeniously improves this aspect. Specifically, each node maintains a small routing table, called *finger table*, which contains at most m entries (i.e., $O(\log N)$ entries). The i-th entry (or the i-th finger) in the finger table of node n records the identifier of the first identifier s (not necessarily corresponds to an existing machine) that succeeds n by at least 2^{i-1} hops on the Chord ring, i.e., we have $s = \text{successor}(n + 2^{i-1})$, where

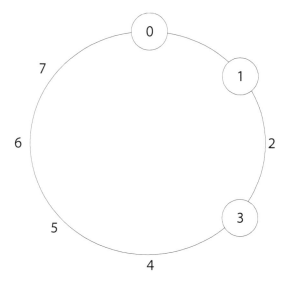

FIGURE 3.1: An example Chord identifier-ring with three nodes currently in the network [Stoica et al., 2001a].

$1 \leq i \leq m$. Furthermore, the i-th entry also includes a half-open interval, starting from s until but not including the identifier of the next finger. The interval of each entry is very useful because it indicates the range of identifiers that particular finger handles (e.g., responsible for storing the data items within the range of identifiers). Most importantly, the i-th entry also contains the identifier q of the successor node (i.e., $q = \text{successor}(s)$). Figure 3.2 shows the finger tables of the three nodes in the Chord ring example with $m = 3$.

To illustrate the data item search process [Balakrishnan et al., 2003] (e.g., a file search in a file sharing system, or a tracker search in a video streaming system), suppose node 3 needs to retrieve a data item with identifier 1. Checking node 3's finger table, 1 is within the interval $[7, 3)$ and the corresponding successor node is 0. Thus, node 3 contacts node 0 to check the latter's finger table in order to locate identifier 1. Eventually, node 0 finds that identifier 1 is handled by node 1, and reports this back to node 3.

Chord can also handle peer dynamics in a robust manner. Specifically, Chord can maintain the following two invariants:

1. Each node's successor (i.e., the first finger) is correctly maintained.

2. For every data item identifier k, it is handled by node with identifier $\text{successor}(k)$.

When a new peer n joins the Chord network, the following three tasks are carried out:

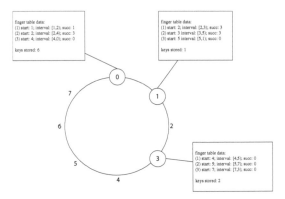

FIGURE 3.2: Finger tables of the three nodes in the Chord ring [Stoica et al., 2001a].

1. Construct the finger table of node n.

2. Update the finger tables of existing nodes to incorporate the existence of n.

3. Notify the higher layer software (e.g., the file sharing application) in order to properly transfer the data items with the associated identifiers which are to be handled by node n.

The reverse of the above tasks need to be done when a node leaves the network.

Figures 3.3 and 3.4 illustrate the situations when node 6 joins and then node 3 leaves.

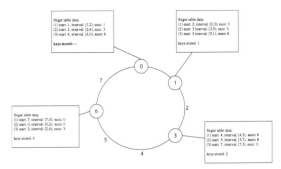

FIGURE 3.3: Updated finger tables of the four nodes in the Chord ring after node 6 joins [Stoica et al., 2001a].

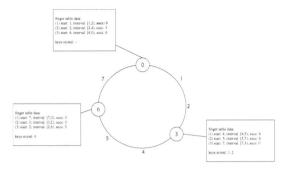

FIGURE 3.4: Updated finger tables of the three nodes in the Chord ring after node 3 leaves [Stoica et al., 2001a].

3.2.2 CAN (Content Addressable Network)

CAN (Content Addressable Network) [Ratnasamy et al., 2001] is another prominent pioneering distributed hash table (DHT) design. Similar to Chord, each CAN participant handles a region of the entire key-value space, called a zone. To support efficient routing, each node also stores information about a small number of adjacent zones in the hash table. Specifically, all the nodes in the network are logically organized into a virtual d-dimensional Cartesian coordinate space. For example, Figure 3.5 depicts a 6-node CAN organized into a 2-D space. Here, node 1 handles the square zone: (2, 2), (3, 2), (3, 3), (2, 3).

With this logical organization, for any data item V with key K, the key K is hashed into a point in the coordinate space. The data item is then stored at the node that handles the zone containing the point. To retrieve a data item, the same hash function is used to locate the zone and request the corresponding node for the item.

To support routing of requests, each node needs to determine and record its neighbors' IP addresses. The "neighbor" notion is intuitive: two nodes (hence two zones) are neighbors if their coordinate spans overlap along exactly $d - 1$ dimensions in the d-dimensional space. As can be seen in Figure 3.5, node 1's neighbors are nodes 2, 3, 4, and 5, but not 6. Routing is then very simple—a node just greedily forwards a message (or request) to the neighbor with coordinates closer (in terms of Cartesian distance) to the destination coordinates. Figure 3.5 shows how node 1 routes a message to the coordinates (x, y) via node 4.

In general, for a d-dimensional space divided into n equal-sized zones, it can be shown that the mean routing path length is $(d/4)(n^{1/d})$. Furthermore, each node in the network needs to maintain $2d$ neighbors. These expressions indicate that the CAN design is highly scalable because per node state is independent of the number of nodes in the system. Even for the path length, the growth rate is only $O(n^{1/d})$. Another important feature of a CAN is its

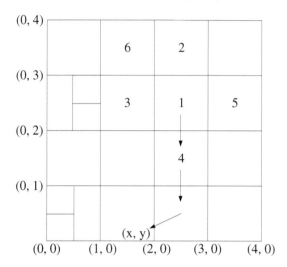

FIGURE 3.5: An example CAN with six nodes organized into a 2-dimensional space in which node 1 routes through 4 to reach a given point (x, y) [Ratnasamy et al., 2001].

robustness—as there are many different paths available for any two points in the d-dimensional space, even if some nodes' neighbors are unavailable (e.g., just departed or crashed), the node can route the message via some other paths.

The process to establish a CAN is also simple and hence robust. Specifically, a node joining the CAN has to carry out the following three steps:

1. Find a node already in the CAN;

2. Find a node whose zone is to be split (so that the new node can take a half); and

3. Notify the neighbors of the split zone to update their neighborhood information.

The first step, i.e., the bootstrapping process, can be implemented in many ways. One common method is to store some default or bootstrap nodes' addresses in some public domain so that the new node can easily obtain them. For the second step, the new node just randomly selects a point in the d-dimensional space and sends a JOIN request to the node that handles the corresponding zone. Notice that this JOIN request is sent to the known bootstrap node which will then route it to the appropriate node handling the selected point. Subsequently, the selected node splits its zone in half and the new node takes one half. The associated keys and data items are then transferred to the new node. Finally, the neighborhood information of the affected nodes is updated.

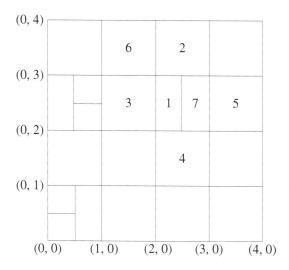

FIGURE 3.6: The updated situation of the CAN after node 7 joins [Ratnasamy et al., 2001].

Figure 3.6 shows the situation after node 7 joins the example CAN. Notice that the members of node 7's neighbor set 1, 2, 4, 5 will update their corresponding neighborhood information.

3.2.3 Other Structured Approaches

Let us first briefly examine two other well-known pioneering DHT approaches: Pastry [Rowstron and Druschel, 2001a] and Tapestry [Zhao et al., 2004]. In the Pastry approach, each node identifier is a 128-bit number picked from a circular key space also. Each peer keeps a routing table having $\log_b N$ rows (here, b is some system-specific integer value). In each row, there are nodes with identifiers matching one prefix more than those of the previous row. Consequently, routing works by matching the identifier in the local routing table for the longest shared prefix with the key. In terms of time-complexity, Pastry DHT systems route a message within $O(\log_b N)$ hops. Tapestry is similar in design as Pastry. Specifically, each node keeps connecting to a set of peers that share common prefixes with its identifier. A Tapestry DHT can also route a message in $O(\log_b N)$ steps.

Fu *et al.* [Fu et al., 2008] proposed an internetworking approach for connecting different DHT networks together, as shown in Figure 3.7. This is an important idea because we can expect a variety of DHTs are being used in practice. Thus, for different P2P system implementations to work together, it is necessary to have such "bridging" mechanisms.

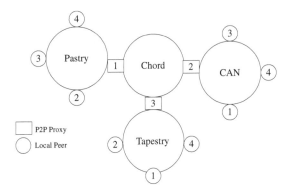

FIGURE 3.7: An internetwork of DHT systems [Fu et al., 2008].

As in a bridge device in a network, the P2P proxy nodes need to support a dual-protocol stack, as shown in Figure 3.8. Obviously, this requires a higher capability for such proxy nodes, similar to the situation of a "Super-Node" in a hierarchical P2P network used in applications such as Skype [Skype, 2009]. To handle the dynamics (i.e., join and leave) of these important proxy nodes, an election mechanism needs to be used [Fu et al., 2008].

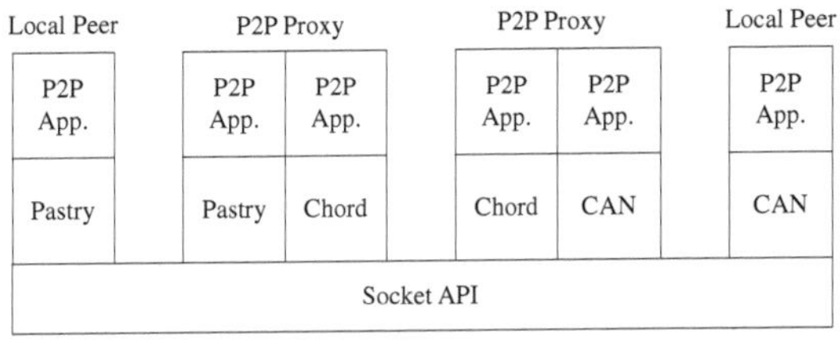

Local Peer	P2P Proxy		P2P Proxy		Local Peer
P2P App.	P2P App.	P2P App.	P2P App.	P2P App.	P2P App.
Pastry	Pastry	Chord	Chord	CAN	CAN
Socket API					

FIGURE 3.8: An example protocol stack supporting the implementation of an internetwork of DHTs [Fu et al., 2008].

Qu et al. [Qu et al., 2009] proposed another interesting P2P internetworking architecture called truncated pyramid, as shown in Figure 3.9. The essence of this network architecture is to interconnect different local P2P overlays (e.g., using Pastry or Chord) by trees. The top overlay, which is the smallest in size, comprises relative more power nodes (e.g., like Super-Nodes) responsible for serving as roots of these trees. Less powerful nodes are then designated roles of internal nodes of the trees. The bulk of other peers, usually forming a much bigger overlay themselves, are designated as leaves. The communications among peers in different levels follow the tree paths. For example, when node 1 wants to communicate with node 13, it can either go through node 11 or

go through nodes 5/6 in its local overlay. To enhance the communication efficiency, "vertical tunnels" can be used—e.g., a top level peer can communicate
directly with a bottom level peer using a tunnel [Qu et al., 2009].

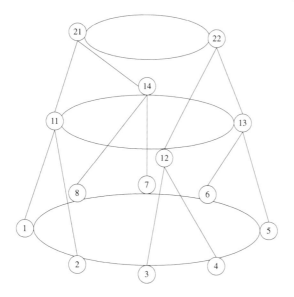

FIGURE 3.9: A truncated pyramid P2P network architecture internetworking several local overlays [Qu et al., 2009].

3.3 Unstructured (Mesh) P2P Systems

As we have seen in Chapter 2, unstructured P2P systems simply involve
random and ad hoc connections among peers [Lv et al., 2002]. Indeed, there
is absolutely no central rule governing the formation of connections among
peers. In a sense, it is truly peer-to-peer in the communication aspects in that
connections are made in a purely autonomous manner. One important feature
of an unstructured network is that as time goes by, those long-lived nodes (i.e.,
nodes that stay on for an extended period of time) usually would have a larger
number of connections due to possibly their resourcefulness (e.g., in terms of
file chunks possessed over time). In some applications (e.g., Skype [Skype,
2009]), such nodes would even be designated as "Super Nodes"—nodes that
have more important responsibilities such as routing, traversal through NATs,
etc.

As to the construction of an unstructured P2P network, there are two approaches. The first one is purely random. The second is autonomous matching assisted by central servers called trackers.

A prominent example of the first approach is the Gnutella [Gnutella Protocol Development, 2009, Ripenau, 2001] file-sharing system. As in any fully decentralized file-sharing system, when a client starts, the very first problem is to discover and locate other active peers. In the original design of the Gnutella protocol, this was based on a flooding mechanism—the starting client broadcasts the so-called "ping" messages over the network. When such a ping message is received by an active Gnutella user, it replies a "pong" message to the starting client. Obviously, a more fundamental question is that to whom should the starting client send the requests in the first place? Many heuristics are used in this bootstrapping process. For example, the starting client can use the list of well-known users that come with the client program. Another scheme is to use a Web cache of actively connected machines.

From a scalability point of view, a drawback of the blind flooding approach is that the volume of traffic generated could be large, even if the maximum hop-count a request message can travel is usually limited to 7. Thus, the notion of "ultra-peer" is introduced in the Gnutella protocol. Specifically, some participating peers are designated as ultra-peers which play the role of "hubs" or "routers" in the Gnutella network. When a new client starts, it actually connects to several (e.g., three) such ultra-peers, each of which could be connected to more than 30 other ultra-peers. Essentially, a user (as a leaf node) sends a request message to its ultra-peers which then forward to its connected ultra-peers. Consequently, with such a more hierarchical network structure, the scope that can be reached by a request message becomes much larger yet the traffic volume generated is limited.

For the second approach, a prominent example is a video streaming system like PPLive [PPLive, 2009, Vu et al., 2010]. Simply put, the general process of a P2P video streaming session is illustrated in Figure 2.4. Initially, the new peer visits the so-called log-in server (i.e., the Web site of the system) to select the channel or movie the user wants to watch. The log-in server then redirects the new peer to a particular tracker server which can furnish a list of peers currently watching the same channel to the new peer. Usually the tracker server just randomly picks a subset of peers to form a list for the new peer. The new peer then selects a subset from this list so as to make connection requests. Such selection is, in current implementations, also based on randomization. After connections are established, buffer maps exchange and video packets downloading can be carried out. This general process is the basis of many well-known P2P video streaming systems such as Joost, SopCast, GridCast, UUSee, etc.

Unstructured P2P networks have many nice characteristics, e.g., low network diameter and, more importantly, robustness against peer dynamics and random node failures. However, the lack of structure also makes it hard to accurately locate data or peers. Thus, many researchers set out to design hybrid network structures in the hope of combining the best of both worlds.

3.4 Hybrid P2P Systems

Ohnishi *et al.* [Ohnishi et al., 2007] proposed an interesting hybrid architecture comprising a DHT ring (e.g., Chord) of unstructured small networks, called load balancing clusters. Specifically, as shown in Figure 3.10, the core network is formed by a set of Super Nodes using a DHT such as Chord. Each Super Node, in turn, manages a small unstructured network of regular nodes. The rationale is that the load of each Super Node can be shared by a dynamic population of regular nodes, depending on the access patterns and traffic conditions. Consider a file sharing application as an example, the following procedure can be implemented.

1. A peer searching for a particular file computes the hash value of the desired file.

2. The search request is then forwarded to a cached Super Node.

3. The receiving Super Node checks the request against its own range of hash values and if the request hash value falls within range, Step (4) is executed; otherwise, Step (5) is executed.

4. The Super Node floods the request within its own load balancing cluster (LBC) so as to locate the file, which is then sent to the requesting peer.

5. The Super Node forwards the request to the next Super Node whose identifier is the closest to the request hash value. Execute Step (3) at this new Super Node.

The maintenance of this semi-structured network architecture entails many load balancing considerations and is discussed in Chapter 4.

Lagesse and Kumar [Lagesse and Kumar, 2007] proposed a hybrid architecture, called Utility Based Clustering Architecture (UBCA), which works by overlaying a clustering mechanism on top of an unstructured P2P network, as shown in Figure 3.11. The main clustering criterion is incentive based—peers form a cluster provided they can derive mutual utility gains. The design goals of UBCA are as follows [Lagesse and Kumar, 2007]:

- Enhance availability and quality of resources;

- Encourage resource sharing in the P2P system;

- Application adaptivity; and

- Maintain underlying system's structure and decentralization.

As can be seen from Figure 3.11, UBCA intercepts the P2P application's requests (e.g., search, insert, delete, etc.) to the underlying P2P network (e.g.,

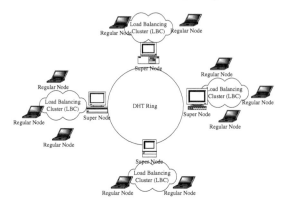

FIGURE 3.10: A semi-structured P2P network architecture [Ohnishi et al., 2007].

a Gnutella random network). Depending upon the clustering status of the peer, UBCA redirects the requests to appropriate peers in the system.

Specifically, UBCA comprises three different layers of components: data, decision logic, and communication, as shown in Figure 3.12. The data layer gathers and maintains information for making clustering decisions. The decision logic layer employs utility functions to evaluate clustering options and select appropriate data items for fetching. The communication layer is responsible for the information exchange among peers.

The information gathered at the data layer is mainly about the qualities and costs of different resources. Utility values can then be computed as qualities less costs. A peer decides to join a cluster if the latter can provide a positive overall utility among all resources concerned. After joining the cluster that provides the maximum aggregate utility, the peer can then use the decision logic layer to select the most important (i.e., with the highest utility gain) resource for retrieval.

With this UBCA architecture, Lagesse and Kumar [Lagesse and Kumar, 2007] showed that the average latency as well as bandwidth utilization are greatly improved.

Hsiao and King [Hsiao and King, 2003] also proposed a very interesting hybrid P2P network architecture for handling mobility of peers. Indeed, the proposed architecture, called Bristle, is designed for managing mobile devices participating in a P2P network. As shown in Figure 3.13, the Bristle architecture consists of two different layers: the stationary layer and the mobile layer. Machines participating in the stationary layer are immobile, i.e., having fixed IP addresses. On the other hand, mobile devices need to join the mobile layer. Consequently, the stationary layer is to be implemented by a DHT, e.g., Chord, while the mobile layer is to be supported by only a random network.

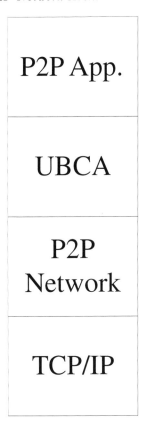

FIGURE 3.11: Protocol architecture of UBCA [Lagesse and Kumar, 2007].

To see how Bristle works, consider a mobile node X trying to communicate with a node Y, whose address is unknown to X. To tackle this problem, X simply sends the message to a known node in the stationary layer. Based on a DHT, the stationary layer can route the message to an appropriate node Z who knows the current address of Y.

3.5 Network Architecture with QoS Provisioning

From a telecom operator's point of view, P2P applications are commonly considered as a threat. Indeed, as the volume of P2P traffic increases (e.g., to something near 90% in some peak hours), P2P applications can lead to significant revenue loss to telecom operators. A further aggravating impact is that both capital and operating expenses also need to be increased in order

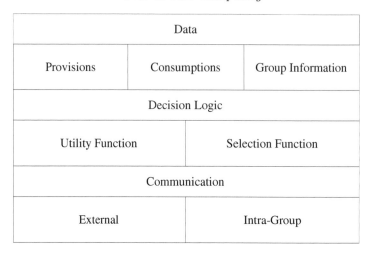

FIGURE 3.12: UBCA components [Lagesse and Kumar, 2007].

to meet the traffic demands. For the lack of a viable P2P business model, telecom operators have yet to reap any tangible profits from the proliferating P2P applications.

Consequently, it is not surprising to see that some telecom operators choose to block P2P traffic. Yet such a "solution" is sometimes considered as a problem by itself because successful and effective blocking of P2P traffic requires high speed detailed packet checking, which obviously adds considerable load to the system and overheads to the users. Furthermore, such blocking of users' traffic inevitably leads to customer complaints.

Thus, some telecom companies start to look at the issue from a more "constructive" angle—trying to find positive ways to work with P2P applications while generating some possibly new revenue streams. Specifically, a natural approach is to investigate whether a telecom operator can design and support a P2P network architecture that can provide carrier-grade QoS guarantees.

Recently, Ma and Zhu [Ma and Zhu, 2008] proposed a novel network architecture that has the following salient features:

- Providing a high scalability with the minimal amount of dedicated network components;

- Supporting a unified approach for customer data management with global accessibility;

- Providing a virtual home environment for customers to enjoy same experiences anywhere;

- Providing QoS guarantees to users; and

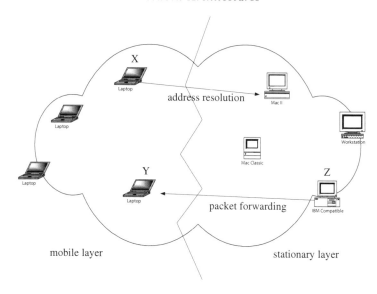

FIGURE 3.13: The Bristle hybrid P2P network architecture [Hsiao and King, 2003].

- Supporting all these services with minimal impact to IP-based bearer network.

The carrier-grade P2P network architecture proposed by Ma and Zhu is depicted in Figure 3.14. The various components are introduced below.

AAA Server. Responsible for authentication, authorization, and accounting.

Charging Mediator. Responsible for collecting charging information (e.g., usage time) and delivering it to the backend billing system.

Bootstrap Server. A default or long-lived server to which a new peer can connect when starting up.

Bootstrap Peer. A more powerful peer elected or designated as a Super Node.

Core Router. IP backbone router.

Edge Router. Ingress or egress router.

Access Router. Responsible for admission control and QoS provisioning.

QoS Coordinator. Responsible for checking and delivering QoS requests to the access network as well as the billing system.

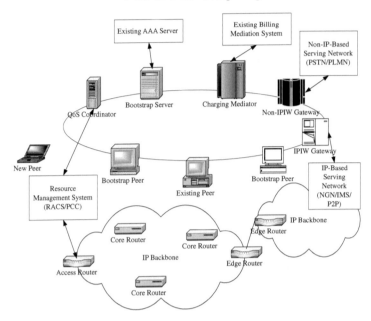

FIGURE 3.14: A carrier-grade P2P network architecture [Ma and Zhu, 2008].

IP-IW Gateway. Carry out necessary protocol functions for internetworking with other P2P networks or IP-based service network.

Non-IP IW Gateway. Carry out necessary protocol functions for internetworking with other non-IP service networks.

There are several key characteristics and requirements of this proposed architecture.

- **Client's Participation in Charging:** In the proposed network architecture, each peer needs to participate in the usage charging process. Specifically, the client program needs to report to the charge mediator about the usage statistics. Here, of course, full cooperation of each peer is a necessity.

- **QoS Initiation by the Peer:** Each peer can dynamically initiate QoS changes which are handled by the QoS mediator.

- **AAA Tasks Handled by Existing Servers:** The tasks of authentication, authorization, and accounting can be handled by the existing AAA servers in the telecom company's network. Thus, there is no need for new investment. Furthermore, current network users can use the existing credential information to access P2P services.

- **Nomadic Support:** Users can have nomadic access to the network as they can obtain their credential information via the Internet.

- **Mobility Support:** Mobility is enabled by using various wireless access technologies (e.g., Wi-Fi).

- **Unique Benefits of the Architecture:** The proposed architecture can potentially combine the best of both telecom services and P2P services because it is designed to provide QoS guaranteed P2P services in a secure and trusted telecom environment. Thus, it is expected that users would agree to pay for such services.

The detailed service provisioning mechanisms are described below.

3.5.1 AAA Tasks

Admission control of peers is handled mainly by the bootstrap server and some bootstrap peers. Specifically, a new peer needs to contact the bootstrap server first. The new peer can then obtain the locations (e.g., IP addresses) of a bootstrap peer. Of course, if the new peer joined the system before, it might have cached the address of a previous bootstrap peer so that it can just directly contact the latter in this instance. In the whole admission process, the new peer works with the assigned bootstrap peer. The bootstrap server together with the bootstrap peer verify the credential information of the new peer.

Mutual authentication, between the new peer and the system, is also handled via the bootstrap server. Upon successful authentication, an authorized token signed by the AAA server is sent to the new peer and incorporated into the JOIN signaling packet to the bootstrap peer. The token is also used in subsequent interactions between the new peer and the system such as getting an authorized QoS level.

3.5.2 Charging

Charging requires cooperation among peers, both the uploading (i.e., serving or supplying) peer and the downloading (i.e., served or consuming) peer. Specifically, the uploading peer requests the downloading peer for some "credit units" which are obtained from the charging system. On the other hand, the charging system also helps the uploading peer to collect, verify, and validate the credits from the downloading peer.

3.5.3 Dynamic QoS

Each participating peer can modify the QoS for any particular session, as illustrated in Figure 3.15.

Specifically, if a peer demands a certain QoS guarantee, it will be charged.

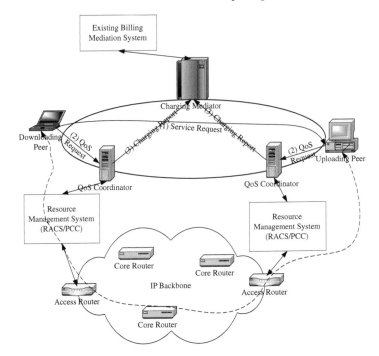

FIGURE 3.15: A dynamic QoS provisioning scenario [Ma and Zhu, 2008].

The charge function of the peer will notify the user about the charge policy. If the user agrees to buy the QoS service, the underlying routers are then configured to provide the required guarantees via proper packet scheduling and other resources provisioning.

3.5.4   Service Brokerage

A brokerage model is used in the proposed network architecture for the provisioning of services among peers. When a peer wants a particular service (e.g., downloading a certain file at a particular data rate), the network acting as a broker attempts to locate such a supplying peer in the system. This is usually accomplished by having some registry servers in the system keeping track of the capabilities and charges required for every participating peer.

3.5.5   Discussion

The P2P network architecture proposed by Ma and Zhu [Ma and Zhu, 2008] described above is a practicable system. However, much of its actual success relies on the cooperation among peers and the proper handling of various security issues. For instance, if some peers collude to cheat the system, the operator (i.e., the telecom company) might take a loss. An example of such collusion is that the peers perform real uploading and downloading of data without notifying the servers so that possibly the "levy" revenues could be lost. Much research still has to be done to improve the design so that such malicious actions can be detected and deterred.

3.6   Video Streaming Network Architecture

Kalogeraki, Delis, and Gunopulos [Kalogeraki et al., 2003] presented a detailed qualitative and quantitative study on different architectures suitable for serving video using a P2P environment. They made some pioneering observations, e.g., that video services would become prominent in P2P systems. Specifically, they delineated the following guidelines for designing and implementing a scalable network architecture for video streaming services.

1. The network architecture and distributed indexing mechanisms should be designed in such a way that efficient retrieval of video data is realized.

2. Queries routing should be carefully designed so that flooding of requests is avoided.

3. Reliability and robustness of the network have to be incorporated so that peer dynamics (i.e., peers come and go) would not disrupt the operation of the video services.

The first network architecture considered by Kalogeraki *et al.* is called the single-index/multiple-index multiple servers (SIMS/MIMS), depicted in Figure 3.16. The key feature of this architecture is that a number of machines are designated as indexing servers, commonly known as trackers nowadays. The indexing servers help participating peers to locate proper serving peers that hold the desired video objects. Each serving peer has an admission control manager component, deciding whether a new connection can be admitted. Furthermore, there is also a QoS manager, taking care of the QoS negotiations (e.g., data rate of uploading) and enforcement.

The second network architecture considered by Kalogeraki *et al.* is called Multiple Independent Indexed Servers (MIIS), depicted in Figure 3.17. Here, the major feature is that each peer keeps partial indexes for the video objects possessed by other peers. The rationale of such a design is to maintain distributed "hooks" in the computing vicinity. Periodically, updates about locally stored new video objects are sent to connected peers so as to maintain a consistent global view of the network.

The third network architecture considered by Kalogeraki *et al.* is called the Fragmented and Multiple Servers (FAMS), depicted in Figure 3.18. This

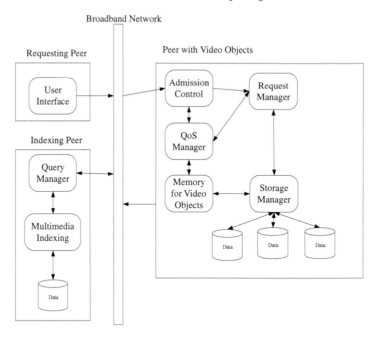

FIGURE 3.16: The SIMS/MIMS network architecture [Kalogeraki et al., 2003].

network architecture is essentially the same as an unstructured P2P network (e.g., Gnutella) in which every peer maintains only its local video objects' information.

Kalogeraki *et al.* then performed simulations to evaluate the performance of the three different architectures. Firstly, they conducted tests to determine the efficiency with which each architecture can reply to a given user request. Specifically, the number of messages needed is counted as the major overhead. It was found that the SIMS/MIMS architecture needs the least number of messages to start the downloading process. On average, the MIIS architecture achieved very similar performance. This is plausibly because as the number of user requests increases, each server/peer adds the locations of more video objects in its local index, as well as attempts to download the most popular video objects. Finally, the overhead expended in the FAMS architecture is the highest and can increase dramatically. This is because of the random search inefficiency—in the worst case, a request has to traverse many nodes before reaching a node that contains the sought video object.

Secondly, they examined the performance of the video object replication in the MIIS architecture. In general, the replication algorithm was effective in that popular video objects were widely replicated while unpopular video objects got a small replication degree.

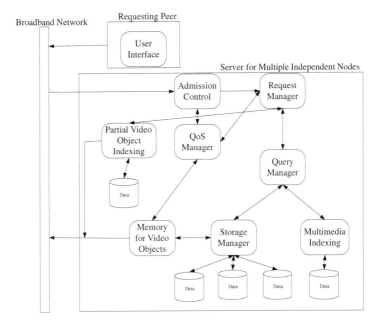

FIGURE 3.17: The MIIS network architecture [Kalogeraki et al., 2003].

Thirdly, the reliability of different architectures was tested by measuring the number of video objects that became unavailable as servers/peers failed. As expected, the FAMS architecture was the most reliable. On the other hand, as time evolved, the reliability of the MIIS architecture gradually improved. This was because the replication algorithm gradually cached more copies of popular video objects.

Fourthly, Kalogeraki *et al.* also investigated the scalability of the architecture as a measure of the number of user requests rejected. In the SIMS/MIMS architecture, the indexing servers/peers can become a performance bottleneck as the number of user requests increases. However, the servers/peers provide QoS guarantees if the requests are admitted. Obviously, in the FAMS architecture, the number of rejected requests is very small, at the expense of having slow connections.

3.7   Case Study: PPLive

The PPLive network is unstructured and is found to be similar to random graphs [Hei et al., 2007a, Vu et al., 2010]. Moreover, in the PPLive mesh network, the average node degree (i.e., number of peers in the neighbor list) is

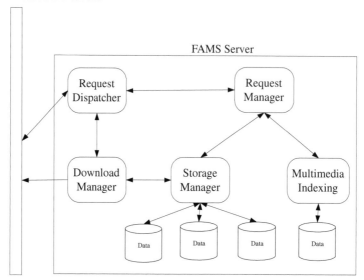

FIGURE 3.18: The FAMS network architecture [Kalogeraki et al., 2003].

independent of the size of the overlay (i.e., the total number of users viewing a particular channel). On the other hand, with reference to a random graph, we can use a metric called Clustering Coefficient (CC) to characterize the organization of the PPLive network. Specifically, CC of a graph is defined as follows: for a random node u and two neighbors v and w chosen randomly from u's neighbor list, CC is the probability that either v is in w's neighbor list or vice versa. It is found [Vu et al., 2010] that the CC of a PPLive network could be quite close to the ratio of average node degree to overlay size, when the overlay size is small. In contrast, when the overlay size increases, the CC value also increases significantly. Furthermore, the availability correlation among PPLive peer pairs is found to be bimodal [Vu et al., 2010]—some pairs have highly correlated availability while others have close to zero correlation. Finally, peer participation in multiple overlays (i.e., downloading/uploading chunks for multiple channels) simultaneously follows a Zipf distribution.

3.8 Summary

Network architecture design has critical impact on the behaviors and performance of a P2P system. At one end of the design spectrum, highly structured network architectures with elegant design and provably good performance, such as CAN and Chord, requires strict compliance to the protocols on the peer's part; otherwise, the structured query and response mechanisms would not work. At the other end, totally unstructured network architectures, such as those manifested by a Gnutella or BitTorrent network, provide high robustness and resilience to peer dynamics, at the expense of having an unpredictable performance. Many efforts are thus proposed to combine the best of both worlds, in the form of some hybrid designs, such as UBCA and Bristle. In media applications such as video and voice, QoS guarantees are very important and highly desirable features, motivating many network architecture designs with QoS provisions. Unfortunately, we have yet to see real life implementations of these research ideas. One plausible reason for the lack of practical implementation of QoS guaranteed architecture is that there are still many unresolved issues. Among them, incentives and security are two of the more crucial ones.

3.9 Review Questions

1. Explain why a Chord DHT system has $O(\log N)$ routing steps.

2. Compare and contrast the CAN and Chord DHT systems.

3. Why is an unstructured network architecture more robust?

4. What are the essential features of a typical hybrid network architecture design?

5. How does a carrier-grade P2P network architecture provide QoS guarantees?

6. Describe the trade-off between a SIMS/MIMS architecture and a FAMS architecture.

Chapter 4

Topology Control

4.1 Introduction

One defining feature of any P2P system is that the composition of the system is highly dynamic and time-varying in nature. In essence, peers join and leave the system at will. Such population dynamics is also sometimes highly unpredictable. A major consequence is that the network topology can change dramatically over time: some high bandwidth links might come and go, or turn into low bandwidth links without notice. Thus, the performance of the participating peers, in terms of downloading/uploading rates, might be adversely affected by the topology changes. To combat these performance degradations that are unavoidable in a practical P2P system, a *topology control* component is necessary.

As P2P networking is all about exchanging information and data efficiently without the help of centralized infrastructure, it is crucial to have an effective topology, one that facilitates fast and robust communications among peers. For example, it would be ideal if peers always form a highly efficient DHT network so that redundancies in file data communication are minimized. However, it is inherently difficult to establish and maintain an effective network topology. This is because in the first place, peers are autonomous and thus, it is difficult to enforce topology establishment and maintenance rules. Secondly, peers come and go and such peer churns also make maintaining an effective topology very difficult. Consequently, *topology control* is a challenging and very important research problem.

Indeed, even if a DHT is used (e.g., Chord), it is mandatory for every participating peer to comply with the maintenance rules such as transfer of data to neighboring peers, which need to accept this responsibility unconditionally. Thus, a major challenge in an effective topology control scheme is that each peer, while rationally selfish, is willing to execute some locally optimizing rules, so as to help maintain an effective global topology.

Topology control is arguably even more important in an unstructured network because in such a network, peer dynamics can easily render an initially good topology very ineffective. Fortunately, in many seemingly random networks, nice global properties emerge, including small-world, power law, etc.

Milgram [Milgram, 1967] pioneered the formal investigation of the "small-world" effects which manifested themselves in the bonding among people with short chains of acquaintances, commonly referred to as six degrees of separation. According to Watts and Strogatz [Watts and Strogatz, 1998], small-world networks exhibit features found in both random and regular network structures. Specifically, the clustering coefficient C_i of a node i is the fraction of all possible edges between adjacent nodes of i that are present:

$$C_i = \frac{D_i}{D_{\max}} \qquad (4.1)$$

where D_i and $D_{\max}$ are the number of adjacent neighbors of node i and the maximum possible number of incident edges of node i.

Moreover, the clustering coefficient C_{net} of a network is the average clustering coefficient of all nodes:

$$C_{\mathrm{net}} = \frac{\sum_i C_i}{|V|} \qquad (4.2)$$

where $|V|$ is the number of nodes in the network.

Consequently, a small-world network typically has a large clustering coefficient as in a regular network but also possesses a small characteristic path length (i.e., the average distance between nodes) as in a random network.

With such salient connectivity features, small-world networks are efficient in facilitating information exchange and dissemination, even for malicious materials such as virus programs. Many P2P applications (e.g., Gnutella [Gnutella Protocol Development, 2009]) are considered to exhibit small-world features.

On the other hand, there is another important notion called *scale-free* or *power-law* [Barabasi and Albert, 1999] network topology. In a scale-free network, the probability that a node is connected to k other nodes is governed by a power-law distribution, $P(k) = k^{-\gamma}$, in which the exponent γ is a value between 2 and 3. Consequently, a large number of nodes have small degrees while a small number of nodes (commonly referred to as hubs) have large degrees. Such a structure, with the hubs, usually exhibits a high resilience against random node failures.

Another critical issue mandating topology control actions is the topology mismatch problem, as illustrated in Figure 4.1. The problem is that while two peers might be logically connected to each other, the "connection" between them can actually involve a highly inefficiency path traversing many other peers. Thus, the key issues here are: (1) detection of inefficient logical connections; and (2) adjustment of topology to better match the underlying physical topology. There is a large body of research addressing this problem.

Many topology control schemes are based on heuristics because optimal topology construction, even in the static case, is an NP-hard problem [Liu et al., 2005b]. Many heuristics use a more or less greedy selection method using parameters such as resourcefulness and connectedness. Specifically, it is intuitively a good idea for a new peer or a disconnected peer to attempt

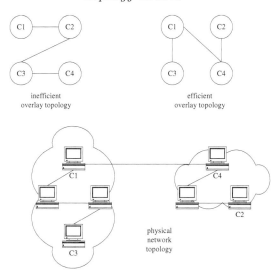

FIGURE 4.1: An illustration of the topology mismatch problem: unnecessary and inefficient overlay connections are made between peers [Liu et al., 2005b].

connecting to a peer with more resources in terms of file data owned, upload bandwidth, etc. Connectivity is also an important consideration because connecting to a high-degree peer presumably allows for better reachability in the network.

However, as mentioned above, if a topology control scheme is designed from a global perspective only, it may not work very well in practice because peers are inherently selfish in the sense that it might not be its interest in optimizing some global performance measures, despite that such "altruistic" actions in fact will benefit the peer in the long run also. Thus, it is important to attack the topology control problem in a fully distributed manner using a game theoretic perspective. Specifically, it might be more practical to formulate some game strategies for each peer so that local (or selfish) optimizations also lead to global performance enhancements.

As elaborated in this chapter below, topology control in a P2P system is still a largely open research topic and has only recently got intensive attention.

We first describe a highly general framework that is adopted by many recent topology control schemes. Specifically, in this framework, the topology control actions are fully distributed—relying only on peers' local actions. As a result, the topology control schemes are highly practical. However, the convergence of the topology control actions might take a long time.

We then survey some recently designed techniques for topology control in a structured P2P network. This is followed by a detailed survey of many interesting unstructured topology control schemes.

We then describe an approach which is based on inference actions performed by each peer in a specialized network-coding-based P2P system. Network coding is a bandwidth efficient packet transmission technique and it allows peers to deduce network bottlenecks.

Finally we also briefly describe a recently proposed topology control technique designed for a wireless P2P system, for which energy efficiency is a prime concern.

4.2   A General Framework for Distributed Topology Control

Singh and Haahr [Singh and Haahr, 2006] suggested using Schelling's model [Schelling, 1971] to dynamically adjust the topology of a P2P network. Specifically, Schelling (an economist) observed that the existence of segregated neighborhoods in the U.S. was neither caused by a central authority nor by the desire of people to stay away from dissimilar people. Instead, the segregation is the cumulative effect of simple actions by individuals who want at least a certain proportion of their neighbors to be similar to themselves.

In Schelling's abstract model [Singh and Haahr, 2006], the world is modeled as a grid. Approximately two-thirds of the cells in the grid are populated by blue or red turtles. The remaining cells are empty. Each cell can host a maximum of one turtle. In the beginning, a random number of blue and red turtles are randomly distributed on the grid. All the turtles desire at least a certain percentage of their neighbors to be of the same color. If a turtle is dissatisfied with its neighbors, it moves to an adjacent empty cell (if available) chosen randomly. This process repeats until all the turtles are satisfied with their neighbors. The resulting segregation is an emergent behavior caused by the desire of the turtles to ensure a certain minimum percentage of their neighbors are the same color as themselves. Schelling's model is thus applicable in a P2P network because each peer lacks a global picture of the network topology. Furthermore, in the model, grouping is maintained even when turtles join or leave the system, which makes it attractive for the dynamic environments of P2P networks.

A similar approach has also recently been suggested by Hariri *et al.* [Hariri et al., 2007] for application level network overlay topology control in a massively multiplayer online game (MMOG). There are many other topology control schemes, such as Auvienen *et al.* [Auvinen et al., 2007], that also fit in this general framework.

4.3 Structured Topology Control

Frey and Murphy [Frey and Murphy, 2008] proposed a detailed mechanism for maintaining a tree network architecture in the presence of peer churns. Specifically, their proposed scheme can limit the maximum node degree, minimize the extent of tree topology changes resulting from peer dynamics, and limit the number of nodes affected by each topology change. Figure 4.2 shows a flow-chart of the topology control algorithm used by each peer whenever it discovers that its parent has departed.

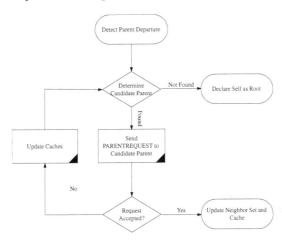

FIGURE 4.2: Algorithm for locating a new parent as a result of peer departures [Frey and Murphy, 2008].

As can be seen, the key step in the topology control algorithm is the identification of a candidate parent. This parent selection step is achieved by using the following strategies.

Regional Strategy. This topology control strategy is designed for repairing the disruption in a localized manner. Specifically, each peer maintains a cache of nearby (in a topological sense) peers, consisting of ancestors as well as siblings. The cache contains a number of ancestor peers starting from the parent continuing toward the root. The cache also contains a set of siblings which are the other children of the parent. When a candidate parent is to be determined, a peer from the cache is randomly selected.

Downstream Strategy. Because repeated selection of ancestors as new parents can lead to high degree values of such peers, violating the design principle of keeping each peer's degree within a reasonable limit, the downstream strategy of topology control works by randomly choosing a candidate parent from the descendants of peers that already have high degree values.

Upstream Strategy. While it is desirable to keep each peer's degree value within a reasonable limit, it is equally important to avoid having a long chain of peers connected in a line. This undesirable situation can happen if leaf nodes are repeatedly selected as candidate parents under the Downstream Strategy. Thus, in this Upstream Strategy topology control action, each peer whose current degree value is lower than a pre-defined lower bound (e.g., in the case of a leaf node), can decline to serve as a candidate parent.

BreakMaxDegree and BreakMinDegree Strategies. While the Downstream and Upstream strategies can help maintain the peer degree values within range, they should not be applied too frequently to the extent that the tree needs to be reconstructed. Thus, the BreakMaxDegree and BreakMinDegree topology control actions are designed for forcing a candidate parent to accept the request if such a parent has declined a previous request due to degree contraints.

Global Strategy. This topology control action is designed to work under a pathological situation where all regional potential parents are unreachable. Specifically, a Global cache is maintained to record some other peers that can be anywhere in the current overlay. Such peers are contacted if all regional strategies fail.

It is obvious that the above topology control actions are by and large independent and therefore, can be combined in any order in consideration. However, in practice it makes more sense to try one strategy (e.g., Regional) first before considering the other (e.g., Global). The above strategies have to be executed in an order such that the tree can be maintained with a low overhead (i.e., relatively fewer nodes are involved) while keeping the shape reasonable (i.e., peer degree values are within range). For example, a sensible order (i.e., a protocol) is: Regional, Upstream, BreakMinDegree, Downstream, Global, and then finally BreakMaxDegree.

One key aspect of tree topology maintenance is to avoid inadvertently forming a cycle. Indeed, each potential candidate parent needs to evaluate whether a request would lead to a cycle in the tree. Indeed, a simple solution is that each peer records precisely its depth (i.e., hop count distance from the root). When a candidate parent is needed, only those with depth values strictly smaller than the current peer are considered. However, while this simple solution is correct, its scalability is poor. For instance, just to maintain correct depth values, the root needs to periodically send a message down the tree so as to update all depth values. Furthermore, after a peer successfully changes its parent, it needs to immediately update the depth values of all its descendants.

Frey and Murphy [Frey and Murphy, 2008] described an ingenious solution to this cycle avoidance problem. Specifically, the key idea is that each peer uses a real number as its depth value, as shown in Figure 4.3.

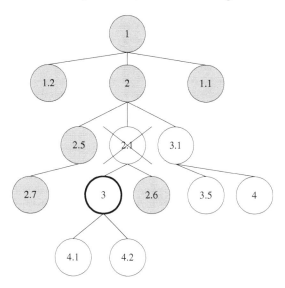

FIGURE 4.3: All shaded nodes are considered as candidate parents for node with depth 3, including the one with depth 3.1 if it reduces its depth value [Frey and Murphy, 2008].

As can be seen from Figure 4.3, the only rule is that each peer's depth value is greater than its parent's. Now, consider the case where a node (e.g., the node with depth value 3) needs to find a new parent. If only integer depth values are allowed, the node with depth value 3.1 is not eligible. However, under a real-valued depth scheme as shown in the figure, such a node (with depth 3.1) can also be a new parent provided it decreases its depth value, say to 2.9 (still larger than its own parent's). Indeed, to guarantee that the tree depth values are consistent, the updating rule is that a peer can only decrease its depth value (while still larger than its parent's) but never increase it. There are two overhead-reducing implications. First, when a peer decreases its depth value, it does not need to coordinate with its descendants. Second, each peer does not need to accurately record the depth value of its parent (which may be updated from time to time without notice). Instead, it only needs to record one correct value (e.g., at the time when the parent is first connected) and can use it afterward. The reason is that the depth value of the parent can only decrease but never increase, so that the peer only needs to make sure that its new depth value (whenever it updates it) is larger than this cached parent depth value.

Figure 4.4 shows a semi-structured P2P network architecture that we discussed in Chapter 3. The major feature of this architecture [Ohnishi et al., 2007] is that each Super Node can make use of its own network (i.e., a load balancing cluster) of regular peers to share the workload. Thus, the topology control actions of this network architecture are mainly related to load balancing concerns.

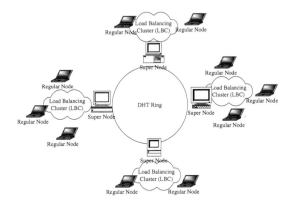

FIGURE 4.4: A semi-structured P2P network architecture [Ohnishi et al., 2007].

Specifically, Ohnishi *et al.* [Ohnishi et al., 2007] considers the following two kinds of load-balanced topology control actions, based on periodic load information exchange among the Super Nodes.

1. A Super Node transfers a regular node in its LBC to another Super Node.

2. A Super Node elevates a regular node in its LBC to be a new Super Node, thereby creating a new LBC.

To initiate one of these topology control actions, each Super Node i periodically checks its average number of object requests (e.g., file searches), denoted as A_i over a certain time period. Two Super Nodes, i and j, can then compare their respective load levels using the ratio A_i/A_j. Specifically, if the ratio is greater than a certain pre-defined threshold, the first kind of transfer from node i to j takes place. On the other hand, if the ratio stays within a certain pre-defined range for all j but the number of regular nodes in i's LBC gets too large, then the second kind of transfer takes place. Here, the number of nodes in each LBC (including the Super Node itself) is governed by:

$$n_r = \beta N_R F_i + \frac{(1 - \beta)N_R}{N_S} \tag{4.3}$$

where N_R is the total number of regular nodes in the system, N_S is the total number of Super Nodes, F_i represents the sum of all ratios at which a request is made for all the objects (e.g., files) included in LBC_i, and β is a system parameter (set to be 1 in [Ohnishi et al., 2007]).

4.4 Unstructured Topology Control

Dale *et al.* [Dale et al., 2008] proposed novel modifications to the tracker servers' peer selection mechanisms in BitTorrent. The objective is to make a BitTorrent network exhibit small-world characteristics. Their proposed modifications are based on a detailed theoretical analysis described below.

To create a small-world network with peers having a known maximum degree, Dale *et al.* first try to maximize the clustering coefficient of a regular graph of the peers. Cliques are considered for this purpose because a network of cliques has a perfect clustering coefficient of 1 and each clique has a maximum node degree. Specifically, a single edge is deleted from each clique and the two incident nodes of the deleted edge are connected to neighboring cliques. This is done uniformly for all cliques in order to maintain regularity. Consequently, a cycle of k identical n-cliques is formed, where n is the maximum node degree and $k = N/n$ (N is the number of nodes).

Now because the network comprises a cycle of identical cliques, it suffices to compute the clustering coefficient of a single n-clique. Recall that each n-clique has two kinds of nodes: (1) $n - 2$ interior nodes connected only to neighbors in the same clique, and (2) two border nodes that connect to neighbor cliques. Notice that the clustering coefficient of a node is equal to the number of triangles including the node. Thus, Dale *et al.* check how many triangles are lost by deleting the single edge so as to connect to neighboring cliques. Obviously, because the interior nodes lose only a single triangle after one edge is deleted, we have:

$$C_i = \frac{\frac{(n-1)(n-2)}{2} - 1}{\frac{(n-1)(n-2)}{2}} = 1 - \frac{2}{(n-1)(n-2)} \tag{4.4}$$

On the other hand, the border nodes lose a triangle for each node that was incident with the missing edge. Because there are $n - 2$ such nodes, the clustering coefficient C_b is given by:

$$C_b = \frac{\frac{(n-1)(n-2)}{2} - (n-2)}{\frac{(n-1)(n-2)}{2}} = 1 - \frac{2}{(n-1)} \tag{4.5}$$

Now considering the average over the n-clique, the clustering coefficient of the entire network is given by:

$$C_G = \frac{(n-2)C_i + 2C_b}{n} = 1 - \frac{6}{n(n-1)} \tag{4.6}$$

Next we consider the diameter and characteristic path length of the network constructed by Dale *et al.*'s method. Observe that we need to traverse three edges to get past a single clique. In the worst case, we need to traverse half way around the cycle of k cliques. Consequently, the maximum diameter is given by: $\frac{3k}{2}$.

For characteristic path length, let us consider only the distances among the interior nodes. Now, for each interior node, there are $n-1$ nodes at distance 1, $2n$ nodes at distance 3 (i.e., one n-clique apart), $2n$ nodes at distance 6, and so on. The sum of the distances for all possible interior nodes is given by:

$$(n-1) + 6n \sum_{j=1}^{\frac{k-1}{2}} j = n - 1 + 3n \frac{k-1}{2}\left(\frac{k-1}{2}+1\right) \tag{4.7}$$

The characteristic path length can then be computed by assuming large values of n:

$$L \approx \frac{n + \frac{3}{4}n(k-1)(k+1)}{nk} = \frac{1 + \frac{3}{4}(k^2-1)}{k} \tag{4.8}$$

To realize the salient features of the network constructed by the above method in a real-life BitTorrent system, Dale *et al.* proposed a simple topology control action to be implemented in a BitTorrent tracker server. Modifying the tracker is a much more realistic approach than modifying all BitTorrent clients. Specifically, the modified tracker assigns an ID to each new peer to designate an n-clique to which it should join. If all the n-cliques are full already, the peer receives a new ID; otherwise, the peer receives the ID of the largest currently unfilled n-clique.

In choosing a list of peers for returning to the new peer for making connections, the n-clique ID of the new peer is considered. Specifically, the tracker first randomly selects a small number of peers from n-cliques other than the one containing the new peer. The rest (i.e., the majority) of the list is formed by randomly selecting peers from the n-clique containing the new peer. The experimental results indicate that such simple modifications can already dramatically increase the extent of clustering, with only a slight increase in network diameter.

In their simulation results [Dale et al., 2008], it is found that the clustering coefficients are increased but not to the extent that the theory predicted. A plausible explanation is that in practice, there is a limit on the number of connections a peer can initiate in a BitTorrent client. This prevents the formation of a complete n-clique.

Kwong and Tsang [Kwong and Tsang, 2008] presented a mathematically sound study on the topology formation and reconstruction schemes in a practical P2P network. Specifically, their key observation is that peers are highly heterogeneous in that they have very disparate bandwidth, storage, processing power capabilities, etc. Thus, it is mandatory to design a suitable protocol for each peer to make connection decisions, i.e., to which peers it should connect. The ultimate objective is to form a network with balanced load across peers and hence, is able to provide stable service quality to users.

To model heterogeneity, Kwong and Tsang [Kwong and Tsang, 2008] use a single parameter η_i for each peer i. Depending on the specific P2P application,

η_i can be thought of representing different system parameters. For instance, in a video streaming application in which the upload bandwidth of each peer is the most critical parameter [Kwong and Tsang, 2008], η_i can be considered as reflecting each peer's upload bandwidth level. On the other hand, in a network coding-based P2P application, CPU processing power is the most important and hence, η_i represents the processing power of each peer i.

In their mathematical model [Kwong and Tsang, 2008], it is assumed that η_i follows a certain probability density function $\rho(\eta)$, which is called the node capacity distribution.

Suppose k_i is the degree of peer i. Presumably every peer wants to connect to a high capacity peer, in the hope that its download performance can be enhanced. On the other hand, it is probably not a good idea to connect to a peer with a high node degree because such a node, while it may be of a high capacity, could also be under a high load. Thus, in Kwong and Tsang's approach [Kwong and Tsang, 2008], two parameters are used—each peer's capacity and node degree—for making connection decisions. Specifically, a probability π_i of a peer i that it is connected by a new peer is given by:

$$\pi_i = \frac{\frac{\eta_i}{k_i}}{\sum_{j \in L(t)} \frac{\eta_i}{k_i}} \tag{4.9}$$

where $L(t)$ is the set of active peers in the system at time t. Intuitively, the probability π_i captures the idea that new peers should heuristically connect to active peers with a large capacity-to-degree ratio, which can in fact be considered as a "normalized" node capacity.

However, it is impractical to maintain global network information in order to compute the probability π_i. Thus, Kwong and Tsang [Kwong and Tsang, 2008] employ the Metropolis-Hastings [Hastings, 1970, Metropolis et al., 1953] to compute the probability in a fully distributed manner as described below.

The P2P network is modeled as a connected graph $G = (V, E)$ with node set $V = \{1, \ldots, n\}$ and undirected edge set $E \subseteq V \times V$. Each edge (i, j) is associated with a transition probability p_{ij}:

$$p_{ij} = \frac{1}{k_i + 1} \min \left\{ 1, \frac{\eta_j k_i (k_i + 1)}{\eta_i k_j (k_j + 1)} \right\} \tag{4.10}$$

and each node i is also associated with $p_{ii} = 1 - \sum_{(i,j) \in E} p_{ij}$.

To compute the edge transition probabilities locally, each node i needs to broadcast its capacity η_i and degree k_i to its neighbors so that the latter can use such information for local computations. These edge transition probabilities are used for random walk.

When a new peer joins the system, it first contacts m active peers in the network, possibly with the help of bootstrap servers. The new peer then dispatches m different walkers to these m nodes. Each walker is associated with a time-to-live (TTL) value denoted as τ, which represents the number of iterations in the Metropolis-Hastings algorithm. Each walker is forwarded from the current node to a neighbor node based on the edge transition probability of the corresponding edge. Each forwarding step leads to decrementing the TTL.

The new peer then connects to the node where the walker stops (i.e., when its TTL reaches 0). If the walker stops at a node which is already connected by that new node, then the walker moves an additional δ hops (in [Kwong and Tsang, 2008], $\delta = 1$). This is repeated until a previously not connected node is found. In practice, this process will not go on indefinitely if the network is large.

Using the above iterative and fully distributed mechanism, the sampled edge transition probability converges to a steady state distribution that is equal to the centralized computation of π_i given above, when $\tau \to \infty$. In [Kwong and Tsang, 2008], it is reported that $\tau = 10$ is already good enough for a network with 50,000 peers.

To handle peer dynamics, a topology reconstruction process is needed. Specifically, in Kwong and Tsang's approach [Kwong and Tsang, 2008], each node i attempts to rebuild r_i (where $r_i \leq 1$) new link(s) whenever it loses a link. Here, obviously r_i is a probability and its usage is detailed as follows. Let k_i^- denote the degree of node i after it has lost a link. Now, a heuristic mechanism is used:

$$r_i = \begin{cases} 1 & k_i^- = 2 \\ r & k_i^- \geq 3 \end{cases} \qquad (4.11)$$

This is called the probabilistic-rebuilding scheme [Kwong and Tsang, 2008]. The threshold of 2 is to heuristically maintain the situation that each node has 3 more links at any point in time. Now, after the probability r_i is determined as shown above, a random walker is dispatched if the decision is to attempt the rebuilding. The same random walk-based node selection process is carried out.

Kwong and Tsang [Kwong and Tsang, 2008] also consider another method called adaptive-rebuilding scheme which works by limiting the degrees of nodes so as to prevent overloading. This is counter-balanced by another heuristic requirement—each node should also try to maintain m links to keep the network's robustness intact. Consequently, in this scheme, the probability r_i is set as:

$$r_i = \frac{m - 1}{k_i^-} \qquad (4.12)$$

Liu *et al.* [Liu et al., 2005b] studied intensively about the overlay topology mismatch problem. They also proposed several heuristics for dynamically adjusting the overlay topology to suit the underlying physical topology. Specifically, their topology control protocol relies heavily on: (1) globally synchronized system clocks; and (2) flooding of slow-connection detector messages.

Let us illustrate their topology control protocol using the example shown earlier in Figure 4.1. As can be seen, C1 and C4 are directly connected in the physical topology. Furthermore, C2 and C4 are in the same subset, as are C1 and C3. Thus, the logical overlay connections C1-C2, C2-C3, and C3-C4 are all inefficient.

In order to discover these inefficient connections, each peer periodically floods some detector messages to their neighbors, which in turn forward them to other nodes, and so on. These detector messages have a controlled scope. For example, their TTL values are set to 2 or 3. On receipt of such a detector message, the TTL value is decremented. More importantly, each message is time-stamped. Thus, when multiple messages originating from the same source are received, the transmission time (i.e., the cost) of the messages can be compared by checking their respective time-stamp values. This is done together with a "probing" action, which works by having a peer send a detector message directly to another peer. The cost of this message is also then noted and compared with the cost of the regular detector messages. Accordingly, some existing connections could be terminated if the newly probed connection generates a lower cost. This heuristic protocol is illustrated in Figure 4.5.

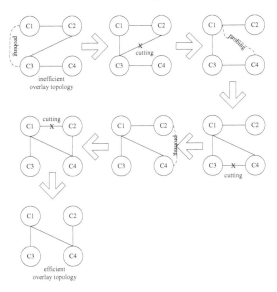

FIGURE 4.5: An example of location-aware topology matching (LTM) [Liu et al., 2005b].

Zhang *et al.* [Zhang et al., 2005a] proposed an interesting scheme for constructing low-diameter overlay networks with power law topologies. In their system, each peer is uniquely identified by a 4-tuple: (IP address, port, network coordinate, capacity). Here, network coordinates are measured using mechanisms such as Vivaldi [Dabek et al., 2004]. Such coordinates are useful for computing the physical "distances" between two peers. Just like many other P2P systems, a new peer i obtains a list of existing peers from a designated bootstrapping server. Specifically, the bootstrapping server selects a list of peers for peer i: $L_i = (LD_i, LR_i)$, where LD_i is a set of several peers with the shortest distances from peer i, and LR_i is a set of randomly selected peers. In Zhang *et al.*'s results, $|LR_i| = |LD_i|$ and $5 \leq |LR_i| + |LD_i| \leq 8$.

Subsequently, peer i sends a probing message to each of the peers on the list. The latter will then reply with their neighbor information. Peer i then assembles all such neighbor information as a candidate list LC_i. For each peer $j \in LC_i$, peer i computes the values of two parameters: the frequency $f_i(j)$ representing the number of times peer j shows up in LC_i thus far, and estimated distance $D(i, j)$ between peer i and j. Furthermore, an important parameter, called normalized distance estimation $d_i(j)$ is determined:

$$d_i(j) = \frac{D(i, j)}{\max_{k \in LC_i} D(i, k)} \tag{4.13}$$

where $0 < d_i(j) \leq 1$.

One interesting way to interpret the significance of these parameters is as follows: LC_i is a sampling of peers in the network; $f_i(j)$ is a sampling of the degree of each candidate j; and $d_i(j)$ is an estimated distance between peer i and j. With such interpretation, another parameter, called connection preference is computed:

$$P_i(j) = \gamma PF_i(j) + (1 - \gamma) PD_i(j) \tag{4.14}$$

where $PD_i(j)$ is the distance preference of peer i connecting to peer j, and thus, serves as the probability that peer i selects peer j as one of its immediate neighbors. Accordingly, it is defined as:

$$PD_i(j) = \frac{\frac{1}{d_i(j)} - \alpha}{\sum_{k \in LC_i} \frac{1}{d_i(k)} - \alpha} \tag{4.15}$$

where $-\infty < \alpha \leq 1$.

By the same token, the degree preference, denoted by $PF_i(j)$, is the probability that peer i selects peer j as one of its immediate neighbors. Here, the more incident edges peer j has in the network, the higher the probability it appears in other peers' candidate lists. Accordingly, degree preference is defined as:

$$PF_i(j) = \frac{f_i(j) - \beta}{\sum_{k \in LC_i} f_i(j) - \beta} \tag{4.16}$$

where $-\infty < \beta \leq 1$.

The parameters, α, β, γ serve as "control knobs" in adjusting the topology of the P2P network. Specifically, larger values of α and γ are more suitable for delay sensitive applications. On the other hand, a larger β and a smaller γ are more suitable for applications that require a better balancing of load. Most importantly, Zhang *et al.* showed that their proposed scheme could indeed generate P2P network structures with power law properties [Zhang et al., 2005a].

4.5   Network-Coding-Based Distributed Topology Control

Jafarisiavoshani *et al.* [Jafarisiavoshani et al., 2007] proposed an interesting topology control algorithm based on network coding [Ahlswede et al., 2000]. In simple terms, using network coding, a source peer sends out to its neighbors a coded packet which is a linear combination (e.g., based on bit-wise XOR operations) of multiple packets that it receives from other neighbors. By attaching also a coding vector to the combined packet, the source allows each receiving neighbor to decode the desired original packet from the combination.

Avalanche [Gkantsidis and Rodriguez, 2005] is a network-coding-based P2P system, in which peers randomly combine their received packets and propagate such linear combinations to their neighbors. A peer receiving a sufficient number of linear combinations solves a system of linear equations (based on the coding vectors retrieved) and retrieves the desired source packets.

Jafarisiavoshani *et al.*'s novel insight [Jafarisiavoshani et al., 2007] is that coding vectors the peers receive from their neighbors can be used to passively infer bottleneck information. This allows individual peers to initiate topology changes to correct problematic connections. In particular, peers, by keeping track of the coding vectors they receive, can detect problems in both the overlay topology and the underlying physical links. The following example illustrates these points.

Consider the example P2P network depicted in Figure 4.6(a) where the edges correspond to logical (overlay network) links. The source S has n packets to distribute to four peers. Peers A, B, and C are directly connected to the source S, and also among themselves with logical links, while peer D is connected to peers A, B, and C. In this overlay network, each peer has a constant degree of three (three neighbors), and there exists three edge-disjoint paths between any pair of peers (in particular, between the source and any other peer).

Now, consider (as shown in Figure 4.6(b)) that the logical links SA, SB, and SC share the bandwidth of the same underlying physical link, which forms a bottleneck between the source and the remaining peers of the network. As a result, let us assume the bandwidth on each of these links is only 1/3 of the bandwidth of the remaining links. Peer D can infer this information by observing the coding vectors it receives from its neighbors A, B, and C.

Specifically, when peer A receives a coded packet from the source, it will forward a linear combination of packets it has already collected to peers B, C, and D. Now each of peers B and C, once they receive the packet from peer A, they also attempt to send a coded packet to peer D. But these packets will not bring new information to peer D, because they are already contained in the combination of coding vectors that peer D has already received. Similarly, when peers B and C receive a new packet from the source, peer D will end

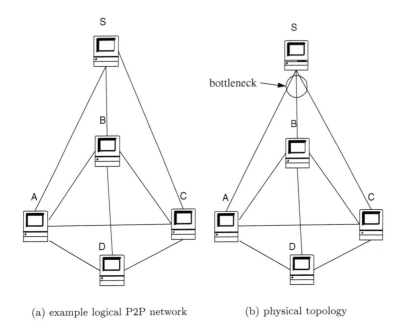

(a) example logical P2P network   (b) physical topology

FIGURE 4.6: Illustrative example [Jafarisiavoshani et al., 2007].

up being offered three coded packets, one from each of its neighbors, and only one of the three will bring to peer D new information. Consequently, peer D can infer from this passively collected information that there is a bottleneck between peers A, B, C and the source, and can thus initiate a connection change.

4.6 Energy Efficient Distributed Topology Control in a Wireless P2P System

Recently, Leung and Kwok [Leung and Kwok, 2008] proposed a comprehensive solution for energy efficient topology control in a hybrid wireless P2P network.

Since such a wireless P2P network is likely to be *ad hoc* in nature, running P2P applications on top of it requires network developers to meet several research challenges. First of all, "Ad hoc P2P" means users are allowed to join and leave freely, and hence, a dynamic credit-system is needed to monitor the behavior of peers so as to monitor "free-riders" [Ge et al., 2003] who do not contribute to the community. Secondly, in such an ad hoc P2P wireless environment, energy efficiency is beyond doubt a crucial factor in the system design due to the fact that mobile wireless devices are inevitably energy limited. Indeed, it is challenging to tackle the energy efficiency problem of mobile devices in a judicious manner and it has intrigued researchers for years, e.g., [Bharghavan et al., 1994, Singh et al., 1998, Singh and Raghavendra, 1998]. This motivates the need for a new energy efficient topology control for effectively supporting P2P file sharing applications.

Topology control, in a traditional sense, comprises two components: neighbor discovery and network organization [Rajaraman, 2002]. In neighbor discovery, one has to detect network nodes in its proximity, and construct a neighbor set in which it could find possible next-hops to establish communication linkages. On the other hand, network organization involves the decision of which communication links to establish with neighboring nodes. Typically, it involves the use of power management schemes such as *sleeping* and *transmit power control*. The former disables some communication links temporarily and the latter adjusts the transmission range.

In summary, traditionally, the objective of topology control is the preservation of network connectivity while improving the efficiency of transmissions. However, Leung and Kwok pointed out that connectivity should be considered at the *application level*. Specifically, their suggested schemes are aimed at achieving an efficient connectivity among mobile devices in order to better serve the file sharing application. Indeed, their idea is that the underlying network layer (or even link layer) connections should be constructed in such a way that the file sharing application's performance is improved. The metric that they use to judge performance is the file success ratio.

Many energy efficient protocols have been proposed in the literature. Nevertheless, previous researchers usually neglect the importance of considering the *difference* of remaining energy levels between *individual nodes*. Indeed, in the pioneering work by Singh [Singh et al., 1998], it only uses metrics for the total energy consumption in the whole path from server-peer (the peer who shares files) to client-peer (the peer who requests files).

The wireless topology control scheme proposed by Leung and Kwok has two components [Leung and Kwok, 2008]. In the first component, called *Adjacency Set Construction*, the topology control system finds out which nodes could be the *next-hop* when a node has to communicate with another node which is n hops away ($n \geq 2$). Specifically, the design rationale is that the construction of neighbor set is related to: (1) contribution levels of different nodes in the P2P file sharing network, (2) the popularity of file resources owned by individual nodes, (3) aggressiveness of the file requesting node, and (4) remaining energy levels of nodes. Our protocol not only takes energy efficiency into consideration but also controls the topology of the file sharing network to introduce fairness.

In the second component, called *Community-Based Asynchronous Wakeup*, the topology control system forms "virtual communities" among mobile users. The term *community* is defined as a set of two or more mobile users that perform in a particular habit. For example, in a music file sharing network, users with similar preference and fans of similar idols are recognized as members of the same "community." Members from the same community follow the same wakeup schedules. The rationale behind this is that file sharing is usually carried out between users with similar interest (this is the usual reason why a sender owns the favorite file that the requester is asking for). Using community formation the topology control system not only increases the chance of getting a file but also allows a group of nodes to sleep and conserve energy when other communities are active.

4.7   Case Study: PPLive

As discussed in Chapter 3, PPLive exhibits a random graph structure [Hei et al., 2007a, Vu et al., 2010]. From a topology control perspective, peers in the PPLive system are pretty autonomous. Connectivity is mainly handled by the usual approach—peers obtain neighbor lists from their respective channel management servers (a.k.a. trackers), and then attempt connecting with peers on the received lists. According to the measurement study conducted by Vu *et al.* [Vu et al., 2010], peers in PPLive incline to choose neighbors that are topologically closer. In other words, the peer selection process seems to be locality aware. Yet perhaps this is due to the fact that a predominantly large population of PPLive users are in China. One the other hand, for the temporal dimension, PPLive peers are found to be quite impatient [Vu et al., 2010]. Specifically, in Vu *et al.*'s performance study [Vu et al., 2010], about 50% of sessions are shorter than 10 minutes. Consider that y is the probability that a node's session length is $10 \times x$ minutes. Vu *et al.* devised a mathematical model of the session time as: $y = ae^{10bx}$, where a and b are some constants (with $a > 0$ and $b < 0$). The degree of peer churn is thus quite high.

4.8   Summary

Topology control is critical for a practical P2P system to deliver good performance in a resilient manner. Specifically, peers must carry out proper and timely topology change actions in response to variations in network connectivity situations due to peers joining or departure. However, such actions have to be localized; otherwise, a great overhead is needed that may actually aggravate the network connectivity changes.

Topology control for wired and wireless P2P systems is still a largely open research topic. Indeed, researchers have not investigated an optimized "interplay" between the application layer, network layer, and even the physical layer (in a wireless setting). In many proposed algorithms, network or physical layer control actions are used (e.g., controlling who is the neighbor). An important next step in topology control research is to propose an efficient "cross-layer" design for P2P systems. Moreover, researchers should come up with a "weighted" combination of the several localized metrics used for setting up new peer connections.

Furthermore, how the existing topology control policies would impact the system evolution subject to different operating conditions (such as existence of non-cooperative peers, selfish peers, or even malicious peers) is an interesting further research topic. Most notably, existence of "free-riders" and "whitewashers" could possibly lower the life-time of the file sharing network significantly because such users would definitely not contribute to the network by not acting as replaying nodes. Furthermore, an even more detrimental situation would be having some malicious users who drop important control messages or fake them, possibly leading to the formation of an inefficient cluster.

4.9   Review Questions

1. Why is topology control important?

2. What are the general techniques used in structured topology control?

3. How do you define the topology mismatch problem?

4. How is network coding useful in topology control?

5. What is the major challenge in wireless topology control?

6. What are the major difficulties in formulating a topology control game?

Chapter 5

Incentives

5.1 Introduction

The highly flexible features of P2P computing such as a dynamic population (users come and go asynchronously at will), dynamic topologies (it is impractical, if not impossible, to enforce a fixed communication structure), and anonymity, come at a significant cost—autonomy, by its very nature, is not always in harmony with tight cooperation. Consequently, inefficient or lack of cooperation could lead to undesirable effects in P2P computing. Among them the most critical one is "free-riding" [Feldman and Chuang, 2005, Ramaswamy and Liu, 2003] behavior. Loosely speaking, free-riding occurs when some users do not follow the presumed altruistic cooperation rules such as sharing files voluntarily, sharing bandwidth voluntarily, or sharing energy voluntarily, so as to benefit the whole community.

Such altruistic sharing actions, presumably, would bring indirect and intangible (and even remote) returns to the users. For instance, if everyone shares files voluntarily, every user would eventually benefit from the high availability of a large and diverse set of selections. Unfortunately, there are some users that do not believe or buy in to such utopia-like concepts and would, then, "rationally" choose to just enjoy the benefits derived from the community, but not contribute their own resources.

To deter or avoid free-riding behaviors, the P2P community has to provide some *incentives*—returns for resource expenditure that are, more often than not, tangible and immediate [Golle et al., 2001]. Such incentives would then motivate an otherwise selfish user to rationally choose to cooperate because such cooperation would bring tangible and immediate benefits. To mention an analogy, in human society, getting pay for our work is a tangible and immediate incentive to motivate us to devote our energy, which could otherwise be spent on other activities. Indeed, it is important for the incentive to be tangible so that a user can perform a cost-benefit analysis—if benefit outweighs cost, the user would then take a cooperative action [Krishnana et al., 2003]. It is also important for the incentive to be immediate (though this is a relative concept) because any resource is associated with an opportunity cost in that if immediate return cannot be obtained from a cooperative action, then the user might want to save the effort for some other private tasks.

To provide incentives in a P2P computing system, there are basically five different classes of techniques.

1. **Payment-Based Mechanisms:** Users taking cooperative actions (e.g., sharing their files voluntarily) would obtain payments in return. The payment may be real monetary units (in cash) or virtual (i.e., some tokens that can be redeemed for other services). Thus, two important components are needed: (1) currency; (2) accounting and clearing mechanism. Obviously, if the currency is in the form of real cash, there is a need for a centralized authority, in the form of an electronic bank, that is external to the P2P system. If the currency is in the form of virtual tokens, then it might be possible to have a peer-to-peer clearing mechanism. In both cases, the major objective is to avoid fraud at the expense of significant overhead. Proper pricing of cooperative actions is also important—over-priced actions would make the system economically inefficient while under-priced actions would not be able to entice cooperation.

2. **Auction-Based Mechanisms:** In some situations, in order to come up with an optimal pricing, auctioning is an effective mechanism. In simple terms, auction involves bidding from the participating users so that the user with the highest bid gets the opportunity to serve (or to be served, depending on context). An important issue in auction based systems is the valuation problem—how much a user should set in the bid? If every user sets a bid higher than its true cost in providing a service, then the recipient of the service would pay too much than is deserved. On the other hand, if the bids are too low, the service providers may suffer. Fortunately, in some form of auctions, proper mechanisms can be constructed to induce bidders to bid at their true costs.

3. **Exchange-Based Mechanisms:** Compared to payment-based and auction-based systems, exchange-(or barter-)based techniques manifest as a purer P2P interaction. Specifically, in an exchange-based environment, a pair of users (or, sometimes, a circular list of users) serve each other in a rendezvous manner. That is, service is exchanged in a synchronous and stateless transaction. For example, a pair of users meet each other and exchange files. After the transaction, the two users can forget about each other in the sense that any future transaction between them is unaffected by the current transaction. This has an important advantage—very little overhead is involved. Most importantly, peers can interact with each other without the need of intervention or mediation by a centralized external entity (e.g., a bank). Furthermore, free-riding is impractical. Of course, the downside is that service discovery and peer selection (according to price and/or quality of service) could be difficult.

4. **Reciprocity-Based Mechanisms:** While pure barter-based interac-

tions are stateless, reciprocity generally refers to stateful and history-based interactions. Specifically, a peer A may serve another peer B at time t_1 and does not get an immediate return. However, the transaction is recorded in some history database (centralized in some external entity or distributed in both A and B). At a later time $t_2 > t_1$, peer B serves peer A, possibly because peer B selects peer A as the client due to the earlier favor from A. That is, as peer A has served peer B before, peer B would give a higher preference to serve peer A. A critical problem is: how to tackle a special form of free-riding behavior, namely the "whitewashing" action (i.e., a user leaves the system and rejoins with a different identity), which enables the free-rider to forget about his/her obligations.

5. **Reputation-Based Mechanisms:** A reputation-based mechanism is a generalized form of reciprocity. Specifically, while a reciprocity record is induced by a pair of peers (or a circular list of more than two peers), a reputation system records a score for each peer based on the assessments made by many peers. Each service provider (or consumer, depending on the application) can then consult the reputation system in order to judge whether it is worthwhile or safe to provide service to a particular client. Reputation-based mechanism is by nature globally accessible and thus, peer selection can be done easily. However, the reputation scores must be securely stored and computed, or otherwise, the scores cannot truly reflect the quality of peers. In some electronic market places such as eBay, the reputation scores are centrally administered. But such an arrangement would again need an external entity and some significant overhead. On the other hand, storing the scores in a distributed manner at the peers would induce problems of fraud. Finally, similar to reciprocity-based mechanisms, whitewashing is a low cost technique employed by selfish users to avoid being identified as low quality users which would be excluded from the system.

The different techniques mentioned above are suitable for different applications. Generally speaking, there are two mainstream applications in P2P environments: sharing of discrete data, and sharing of continuous data. Examples of the former include file sharing systems (e.g., Napster), data sharing systems (e.g., sharing of financial or weather reports), etc. A notable example of the latter is P2P video streaming. Indeed, there is an important difference between file sharing and media streaming systems. In the former, a user needs to wait until a file (or a discrete unit of shared information) is completely received before it can be consumed or used. Thus, there could be a significant delay between service request and judgement of service quality. In an extreme case, a user may not discover that a shared file is indeed the one requested or just a piece of junk. By contrast, in a media streaming application, a user would quickly discover if the received information is good enough. The quality of service (QoS) metric used is also different in these two different applications. In a file sharing application, the most important metrics are downloading time and the integrity of the received files. In a media streaming application, the more crucial performance parameters are the various playback quality metrics such as jitter, frame-rate, resolution, etc. Furthermore, the incentive techniques surveyed in this chapter subsume the underlying P2P network topology. Specifically, in most proposed systems, the communication message exchange mechanism is not explicitly modeled.

Currently, the majority of P2P systems are implemented over the Internet. However, wireless P2P systems are also proliferating. While most of the incentive techniques designed for a wired environment could be applicable in a wireless system, the wireless connectivity is by itself an important bootstrap sharing problem. Indeed, on the Internet, users seldom pay attention to the connectivity issue because a user can be reached (or can reach) any other Internet user without noticeable effort. The only concern about communication is the uploading or downloading bandwidth consumption. In a wireless environment, however, the mere action of sending a request message from a client peer to a server peer would probably need several intermediate peers to help do the message forwarding because the server and client peers may be out of each other's transmission range. Consequently, incentives have to be provided to encourage such forwarding actions.

In this chapter, we survey state-of-the-art solutions proposed for tackling the incentive issues in various different P2P resource sharing systems. In Section 5.2, we describe approaches designed for providing incentives in Internet-based P2P networks. We discuss both file sharing and media streaming applications. In Section 5.3, we describe solutions suggested for wireless P2P systems. We then provide some of our interpretations and suggestions in Section 5.4. We summarize this chapter by providing some remarks in Section 5.6.

5.2   Incentive Issues in P2P Systems on the Internet

5.2.1   File Sharing Systems

In a file sharing system, users would like to retrieve files from other users, and would expect other users to do the same. Thus, each user would need to expend two different forms of resource:

- Storage: Each user has to set aside some storage space to keep files that may be needed by other users, even though such files may not be useful to the user itself.

- Bandwidth: Each user has to devote some of its outbound bandwidth for uploading requested files to other users.

Users usually perform file selection (and hence, peer selection) with the help of some directory system which may or may not be fully distributed. For example, in Napster [Napster, 2009], the directory is centralized.

Using such a sharing model, the most obvious form of free-riding behavior is that a selfish user just keeps on retrieving files from others but refuses to share its collections (and thus, no need to expend any outbound bandwidth for file uploading). Interestingly enough, in an empirical study using the Maze file sharing system [Maze, 2006] performed by Yang *et al.* [Yang et al., 2005], it is found that the more direct indicator of free-riding behaviors is the online time of a user. Specifically, the online time of a selfish user in a P2P file sharing network is on average only one-third of that of a cooperative peer.

In this section, we first briefly overview a contemporary file sharing system called BitTorrent [Cohen, 2003]. We then survey techniques suggested for various other P2P file sharing networks.

5.2.1.1 BitTorrent

BitTorrent [Cohen, 2003] is by far one of the most successful P2P file sharing systems. A key feature in BitTorrent is that each shared file is divided into pieces (of size 256KB each), which are usually stored in multiple different peers. Thus, for any peer in need of a shared file, parallel downloading can take place in that the requesting peer can use multiple TCP connections to obtain different pieces of the file from several distinct peers. This feature is highly effective because the uploading burden is shared among multiple peers and the network can scale to a large size. Closely related to this parallel downloading mechanism is the incentive component used in BitTorrent. Specifically, each uploading peer selects up to four requesting peers in making uploading connections. The selection priority is based on descending order of downloading rates from the requesting peers. That is, the uploading peer selects four requesting peers that have the highest downloading rates. Here, downloading rate refers to the data rate that is used by a requesting peer in sending out pieces of some other file. Thus, the rationale of this scheme is to provide incentive for each participating peer to increase the data rate used in sending out file data (i.e., uploading, or, in BitTorrent's term, *unchoking*). There are other related mechanisms (e.g., optimistic unchoking), which are described in detail in [Cohen, 2003, Qiu and Srikant, 2004].

Qiu and Srikant [Qiu and Srikant, 2004] performed an indepth analysis of BitTorrent's incentive mechanism. By using an intricate and accurate model, it is shown that a Nash equilibrium exists in the upload/download game in BitTorrent. At the equilibrium, each peer sets its uploading data rate to be its physical maximum uploading rate (i.e., each peer is fully cooperative). On the other hand, due to the usage of the optimistic unchoking mechanism (a fifth requesting peer is randomly selected in the uploading process, for details, see [Cohen, 2003, Qiu and Srikant, 2004]), a free-rider can potentially achieve 20% of the possible maximum downloading rate. This theoretical result conforms nicely with the simulation findings by Jun and Ahamad [Jun and Ahamad, 2005] who observed that in BitTorrent, free-riders are not penalized adequately while contributors are not rewarded sufficiently.

5.2.1.2 Hierarchical P2P Systems

In some situations, the P2P network may be structured in a hierarchical manner so that some specialized machines (called "super-peers") can take up a more important role for handling resource management tasks such as request forwarding and routing, directory listing, etc.

Singh *et al.* [Singh et al., 2003] proposed a super-peer-based scheme. Specifically, Singh *et al.* studied the impacts of super-peers in a P2P file sharing network. Simply put, a super-peer is a special network node that serves as a hub to provide file indexing service to other nodes. The problem is that there is a lack of incentive for a participating node to act as a super-peer because any node can simply join an existing super-peer to obtain good performance. Singh *et al.* observed that some entities external to the P2P system, such as an Internet Service Provider (ISP) or a content publisher, have business driven incentives for designating some nodes to act as super-peers, and for enhancing the capabilities of super-peers. Specifically, a super-peer can cache meta-data only instead of the files themselves. As such, the cost of acting as a super-peer is lower. Furthermore, with properly designed meta-data, a super-peer can support value-added search commands (e.g., topic-based search of files).

With such facilities incorporated in each super-peer, a hierarchical P2P file sharing network is then much more efficient than a flat P2P system which relies on request-flooding. Consequently, all nodes in the system have the incentive in maintaining such a P2P file sharing system. Simulation results also indicate that the proposed super-peer scheme is effective.

5.2.1.3 Payment-Based Systems

Hauscheer *et al.* [Hausheer et al., 2003] suggested a token-based accounting system that is generic and can support different pricing schemes for charging peers in file sharing. The proposed system is depicted in Figure 5.1. The system mandates that each user has a permanent ID authenticated by a certification authority. Each peer has a token account keeping track of the current amount of tokens, which are classified as local and foreign. A peer can spend its local tokens for accessing remote files. The file owner treats such tokens as foreign tokens, which cannot be spent but need to be exchanged with super-peers for new local tokens. Each token has a unique ID so that it cannot be spent multiple times.

At the beginning of each file sharing transaction, the file consumer tells the file owner about which tokens it intends to spend. The file owner then checks against the file consumer's account kept at the file owner's machine. If the tokens specified are valid (i.e., they have not been spent before), then the file consumer can send the tokens in an unsigned manner to the file owner.

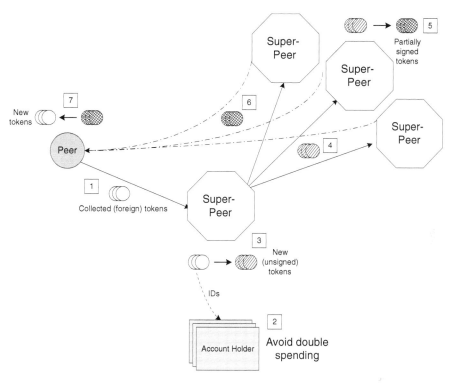

FIGURE 5.1: A super-peer-based token accounting system for P2P file sharing [Hausheer et al., 2003].

Upon receipt of these unsigned tokens, the file owner provides the requested files to the file consumer. When the files are successfully received, the file consumer sends the signed version of the tokens to the file owner. In this manner, Hauscheer *et al.* argued that there is no incentive for the peers to cheat.

Yang and Garcia-Molina [Yang and Garcia-Molina, 2003] proposed the PPay micropayment system in which each peer can buy a coin from a broker. The peer then becomes the "owner" of the coin and can spend it to some other peer. An important feature is that even after the coin is spent, the original owner still has the responsibility to check the subsequent usage of the coin. For example, suppose A is the owner of a coin which is spent to B. If B wants to spend the coin in turn to C, the original owner A needs to check whether such a transaction is valid (e.g., to avoid double spending of the same coin). If A is offline (e.g., temporarily departed the P2P system), then the broker is responsible to perform such checking.

Although the PPay system described above is a useful tool for supporting P2P sharing, Jia *et al.* [Jia et al., 2005] observed that PPay can be further improved. Specifically, Jia *et al.* proposed a new micropayment system, called CPay (an improved version of PPay), which has one significant new feature. The new feature is that the broker judiciously selects the most appropriate peer to be the owner of a coin. Specifically, the owner of a coin should be one that is expected to stay in the system for a long period of time. Thus, the broker's potential burden of checking coin owners' transactions can be considerably reduced.

Figueiredo *et al.* [Figueiredo et al., 2005] also considered a payment-based system to entice cooperation among peers. Specifically, each peer requiring message forwarding service from other peers needs to pay real money to these peers. Thus, as a peer stays in the P2P network and provides forwarding service to other peers, it can gain money. Indeed, such a monetary gain represents an incentive to enhance the availability of a peer in the network.

Saito [Saito, 2003] proposed an Internet-based electronic currency called *i*-WAT, which can be used by users in a P2P system for "purchasing" services. Each *i*-WAT message is an electronic ticket signed using OpenPGP. Saito *et al.* [Saito et al., 2005] then extended the *i*-WAT system by adding a new feature called "multiplication over time," which means that a requesting peer's debt (in terms of *i*-WAT units) increases over time. This feature then encourages service providing peers to stay in the system for a longer period of time so as to defer the redemption of *i*-WAT tickets, thereby increasing the gains from the requesting peers.

5.2.1.4 Cost of Sharing

Varian [Varian, 2003] reported a simple but insightful analytical study on disincentives in P2P sharing. Table 5.1 lists the notation used in Varian's analysis.

TABLE 5.1: Notation used in Varian's analysis on disincentives for P2P sharing.

Symbol	Definition
p	unit price
v	value of the item as perceived by each peer
n	number of peers in the system
D	total cost of producing all the items
$d = \frac{D}{n}$	average development cost
k	number of peers in each group that share an item
t	sharing cost incurred by each member of a group
c	cost imposed by the central authority to those peers who participate in sharing
π	profit derived by the central authority

Note: From Varian, 2003.

In Varian's model, a single item (e.g., a single music file) is considered and the system is homogeneous in that all peers have the same valuation on the item. There is an external central authority (e.g., the original producer of the music) that has the incentive to discourage sharing among peers in the system. The issue is then how the central authority can introduce proper disincentives into the system. Firstly, observe that for viability in producing the item, we have:

$$v - \frac{p}{k} - t - c \geq 0 \qquad (5.1)$$

$$p\frac{n}{k} \geq D \qquad (5.2)$$

Specifically, the above equations are to ensure that value is no smaller than the cost.

The equilibrium (where the item is just viable to be produced) price and profit are then given by:

$$p = (v - t - c)k \qquad (5.3)$$

$$\pi = (v - t - c)kn - D \qquad (5.4)$$

We can see that profit is, counterintuitively, decreasing in c. An interpretation is that c is not large enough to discourage sharing and price has to be cut for compensation. As an analogy, consider that c represents some copy protection mechanism which merely brings inconvenience to customers but cannot discourage sharing. To compensate for the inconvenience, the price has to be reduced.

On the other hand, for a given value of c, suppose the price is set in such a way that it is marginally unattractive to share. That is, we have:

$$\frac{p}{k} + t + c \geq p \qquad (5.5)$$

This in turn implies that:

$$p = \frac{k}{k-1}(t + c) \qquad (5.6)$$

Now, as the maximum practical value of p is v, we have:

$$c = v - \frac{k-1}{k}t \qquad (5.7)$$

Yu and Singh [Yu and Singh, 2003] also investigated the issue of proper pricing in the presence of free-riders in a P2P system. A referral-based system is considered in that each peer can either answer a remote query directly (e.g., serving a requested file) or reply with a referral (e.g., pointing to a different peer who may be able to serve the requested file). A requester (i.e., file consumer) needs to pay for both a referral or a direct answer. Each peer keeps track of reserve prices of potential referrals and direct answers from any other peer. These reserve prices are updated dynamically based on transaction experiences in that a satisfactory transaction leads to an increase in the reserve prices while an unsatisfactory one leads to a decrease. From a seller's point of view, these prices are also exponentially decreased as time goes by. Under this model, simulation results indicate that a free-rider will quickly deplete its budget. On the other hand, the price of a direct answer is found to be much higher than that of a referral.

Courcoubetis and Weber [Courcoubetis and Weber, 2006] recently reported an indepth analysis of the cost in sharing in a P2P system. In their study, a P2P sharing system is modeled as a community with an excludable public good. Furthermore, the public good is assumed to be nonrivalrous, meaning that a user's consumption of the public good does not decrease the value of the good. Such a model is suitable for a P2P file sharing network, where the excludable public good is the availability of shared files. The model is also considered as suitable for a P2P wireless LAN environment, where the excludable public good is the common wireless channel. With their detailed modeling and analysis, an important conclusion is derived: each peer only needs to pay a fixed contribution, in terms of service provisioning (e.g., a certain fixed number of distinct files to be shared by other peers), in order to make the system viable. Such a fixed contribution is to be computed by some external administrative authority (called a "social planner") by using the statistical distribution of the peers' valuations of the public good.

5.2.1.5   Reciprocity and Reputation-Based Systems

Feldman *et al.* [Feldman et al., 2004a] suggested an integrated incentive mechanism for effectively deterring (or penalizing) free-riders using a reciprocity-based approach. Specifically, the proposed integrated mechanism has three core components: discriminating server selection, maxflow-based subjective reputation computation, and adaptive stranger policies.

In the discriminating server selection component, each peer is assumed to have a private history of transactions with other peers. Thus, when a file sharing request is initiated, the peer can select a server (i.e., a file owner) from the private history. However, in any practical P2P sharing network, we can expect a high turnover rate of participation. That is, a peer may only be present in the system for a short time. Thus, when a request needs to be served, such a departed peer would not be able to help if it is selected. To mitigate this problem, a shared history is to be implemented. That is, each peer is able to select a server from a list of global transactions (i.e., not just restricted to those involving the current requesting peer). A practical method of implementing shared history is to use a distributed hash table-(DHT-) based overlay networking storage system [Stoica et al., 2001b]. Specifically, a DHT is an effective data structure to support fast look-up of data locations.

A problem in turn induced by the shared history facility is that collusion among non-cooperative users may take place. Specifically, the non-cooperative users may give each other a high reputation value (e.g., possibly by reporting bogus prior transaction records). To tackle this problem, Feldman *et al.* suggested a graph theoretic technique. To illustrate, consider the reputation graph shown in Figure 5.2. Here, each node in the graph represents a peer (C denotes a colluder) and each directed edge represents the perceived reputation value (i.e., the reputation value of the node incident by the edge as perceived by the node originating the edge). We can see that the colluders give each other a high reputation value. On the other hand, a contributing peer (e.g., the top node) gives a reputation value of 0 to each colluder because the contributing peer does not have any prior successful transaction carried out with a colluder. With this graph, we can apply the maxflow algorithm to compute the reputation value of a destination peer as perceived by a source peer. For instance, peer B's (the destination) perceived reputation value with respect to peer A (the source) is 0 despite that many colluders give a high reputation value to B.

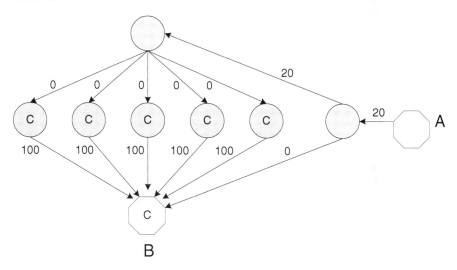

FIGURE 5.2: A graph depicting the perceived reputation values among peers (C denotes a colluder) [Feldman et al., 2004a].

Finally, an adaptive stranger policy is proposed to deal with whitewashing. Instead of always penalizing a new user (which would discourage expansion of the P2P network), the proposed policy requires that each existing peer, before deciding whether to do a sharing transaction with a new user, computes a ratio of amount of services provided to amount of services consumed by a new user. If this ratio is great than or equal to 1, then the existing peer will work with the new user. On the other hand, if the ratio is smaller than 1, then the ratio is treated as a probability of working with this new user.

Sun and Garcia-Molina [Sun and Garcia-Molina, 2004] suggested an incentive system called Selfish Link-based InCentive (SLIC), which is based on pairwise reputation values. Specifically, any peer u maintains a reputation value $W(u, v)$ for each of its neighbor peer v, where the reputation value is normalized such that $0 \leq W(u, v) \leq 1$. Here, "neighbor" means a peer v currently having a logical connection with u and thus, such a peer v can potentially request for service from u. With these reputation values, the peer u can then allocate the uploading bandwidth to any requesting neighbor peer v with a value of $W(u, v)/\sum_i W(u, i)$. The reputation value $W(u, v)$ is updated periodically based on an exponential averaging method.

Under this model, Sun and Garcia-Molina [Sun and Garcia-Molina, 2004] observed that each peer has the incentive to do some or all of the following, in order to increase its reputation values as perceived by other peers (and hence, enjoy a better quality of service).

- Sharing out more file data;

- Connecting to more peers (to increase the opportunities for serving others);

- Increasing its total uploading capacity.

5.2.1.6 Penalty-Based Approaches

Feldman *et al.* [Feldman et al., 2004b] also investigated disincentive mechanisms that can discourage free-riding. Specifically, they considered various possible penalty schemes in deterring free-riders. A simple model is used. At the core of the model, each user i in the P2P sharing network is characterized by a positive real-valued *type* variable, denoted as t_i. Another key feature of the model is that the cost of contributing is equal to the reciprocal of the current percentage of contributors, which is denoted as x. Thus, for any rational user with type t_i, the user will choose to contribute if $1/x < t_i$ and free-ride if $1/x \geq t_i$.

Furthermore, the benefit each user derived from the P2P network is assumed to be of the form αx^β, where $\beta \leq 1$ and $\alpha > 0$. With this benefit function, the system performance is defined as the difference between the average benefit and the average contribution cost. Specifically, system performance is equal to: $\alpha x^\beta - 1$.

Even with the simplistic model described above, Feldman *et al.* provided several interesting conclusions. Firstly, it is found that excluding low type users can improve system performance only if the average type is low and α is large enough. Unfortunately, exclusion is impractical because a user's type is private and thus, cannot be determined accurately by other peers. It is then assumed that free-riding behaviors are observable (i.e., free-riders can be identified). Such free-riders are then subject to a reduction in quality of service. Quantitatively, the benefit received by a free-rider is reduced by a factor of $(1-p)$, where $0 < p \leq 1$. A simple implementation of this penalty is to exclude a free-rider with a probability of p. The second interesting conclusion is that the penalty mechanism is effective in deterring free-riders when the penalty is higher than the contribution cost. In quantitative terms, the condition is that $p > 1/\alpha$. Finally, another interesting conclusion is that for a sufficiently heavy penalty, no social cost is incurred because every user will contribute (i.e., choose not to be a free-rider) so that optimal system performance is achieved. In particular, to deal with the whitewashing problem, the analysis suggests that every new user is imposed a fixed penalty. Essentially, this is similar to the case in the eBay system where every new user has a zero reputation and thus, will less likely be selected by other users in commercial transactions. However, this is in sharp contrast to the adaptive stranger policies suggested also by Feldman *et al.* in another study [Feldman et al., 2004a] that we have described earlier.

5.2.1.7 Game Theoretic Modeling

Ranganathan *et al.* [Ranganathan et al., 2003] proposed and evaluated three schemes induced by the Multi-Person Prisoner's Dilemma (MPD) [Osborne, 2004, Schelling, 1978]. The basic Prisoner's Dilemma game models the situation where two competitors are both better off if they cooperate than when they do not. However, without communication, the unfortunate stable state is that both competitors would choose not to cooperate. An MPD is a generalization of the basic PD. Specifically, the key features of the MPD framework can be briefly summarized as follows:

- The MPD game is symmetric in that each of n players has the same actions, payoffs, and preferences.

- Any player's payoff is higher if other players choose some particular actions (e.g., "quiet" instead of "fink").

The MPD framework is used for modeling P2P file sharing as follows. There are n users in the system, each of which has a distinct file that can be either shared or kept only to the owner. The system is homogeneous in that all files have the same size and same degree of popularity. Now, the potential benefit gained by each user is the access of other users' files. The cost involved is the bandwidth used for serving other users' requests. With this simple model, it can be shown that the system has a unique Nash equilibrium in which no user wants to share. Obviously, this equilibrium is sub-optimal (both at the individual level and at a system-wide level) in that each user could obtain a higher payoff (i.e., a higher value of net benefit) if all users choose to share their files.

Motivated by the MPD modeling, Ranganathan *et al.* proposed three incentive schemes:

- **Token Exchange:** This is a payment-based scheme because each file consumer has to give a token to the file owner in the sharing process. Each user is given the same number of tokens initially and each file has the same fixed price.

- **Peer-Approved:** This is a reputation-based scheme in that each user is associated with a rating which is computed using metrics such as the number of requests successfully served by the user. A user can download files from any owner who has a lower or the same rating. Thus, to gain access to more files in the system, a user has to actively provide service to other users so as to increase the rating.

- **Service Quality:** This is also a reputation-based scheme similar to Peer-Approved. The major difference is that a file owner provides differentiated service qualities to users with different ratings.

Theoretical analysis [Ranganathan et al., 2003] indicates that the Peer-Approved policy with a logarithmic benefit function (in terms of number of accessible files) can lead to the optimal equilibrium where every user contributes fully to the system. Simulation results also suggest that Peer-Approved generates performance (in terms of total number of files shared) comparable to that of Token Exchange, which entails a higher difficulty in practical implementation as it requires a payment system.

Becker and Clement [Becker and Clement, 2004] also suggested an interesting analysis of the sharing behaviors using variants of the classical 2-player Prisoner's Dilemma. Specifically, the P2P file sharing process is divided into three different stages: introduction, growth, and settlement. In the introduction stage, the P2P network usually consists of just a few altruistic users who are eager to make the network viable. Thus, sharing of files is a trusted social norm. The payoffs of the two possible actions (supply files or not supply files) are depicted in Figure 5.3. Here, we have the payoffs ranking as: $R > T > S > P$ (note: T: Temptation, R: Reward, S: Sucker, P: Punishment). Consequently, the Nash equilibrium profile is: (Supply, Supply). Notice that the payoffs ranking in the original Prisoner's Dilemma is: $T > R > P > S$, and as such, the Nash equilibrium is the action profile in the lower right corner of the table.

Player 1 \ Player 2	Action: Supply g_2^1	Action: No Supply g_2^2
Action: Supply g_1^1	R R	T S
Action: No Supply g_1^2	S T	P P

FIGURE 5.3: Payoff table in the introduction stage [Becker and Clement, 2004].

In the growth stage, we can expect that more and more non-cooperative users join the network. For these users, the payoffs ranking becomes: $T > R >$

$P > S$, which is the same as the original Prisoner's Dilemma. Thus, the Nash equilibrium for such users occurs at the profile: (No Supply, No Supply). As the P2P network progresses to the mature stage (i.e., the size of the network becomes stabilized), we can expect that a majority of users are neither fully altruistic nor fully non-cooperative. For these users, the payoffs ranking is: $R > T > P > S$. As a result, the payoff matrix is depicted in Figure 5.4. As can be seen, there are two equally probable Nash equilibria: (Supply, Supply) and (No Supply, No Supply). Consequently, whether or not the P2P network is viable or efficient depends on the relative proportions of users in these two equilibria. Results obtained in empirical studies [Becker and Clement, 2004] using real P2P networks conform quite well to the simple analysis described above.

Player 2 / Player 1	Action: Supply g_2^1	Action: No Supply g_2^2
Action: Supply g_1^1	R $\quad R$	$\quad T$ S
Action: No Supply g_1^2	$\quad S$ T	$\quad P$ P

FIGURE 5.4: Payoff table in the settlement stage [Becker and Clement, 2004].

Ma *et al.* [Ma et al., 2004a, Ma et al., 2004b] suggested an analytically sound incentive mechanism based on a fair bandwidth allocation algorithm. Indeed, the key idea is to model the P2P sharing as a bandwidth allocation problem. Specifically, the model is shown in Figure 5.5. Here, multiple file requesting peers compete for uploading bandwidth of a source peer. Each requesting peer i sends a bidding message b_i to the source peer N_S. The source peer then divides its total uploading bandwidth W_S into portions of x_i for the peers. However, due to network problems such as congestion, each peer i may receive an actual uploading bandwidth of x'_i which is smaller than x_i.

Each bidding message b_i is the requested amount of bandwidth. Thus, we have $x_i \leq b_i$. To achieve a fair allocation, the source peer uses the contribution level C_i of each competing peer i to determine an appropriate value of x_i. Ma *et al.* [Ma et al., 2004a, Ma et al., 2004b] described several allocation algorithms with different complexities and considerations: simplistic equal sharing, max-min fair allocation, incentive-based max-min fair allocation, utility-based max-min fair allocation, and incentive with utility-based max-min fair allocation. The last algorithm is the most comprehensive and effective. It works by solving the following optimization problem:

$$\max \sum_{i=1}^{N} C_i \log(\frac{x_i}{b_i} + 1) \qquad (5.8)$$

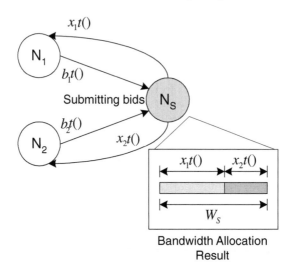

FIGURE 5.5: Two file requesting peers (N_1 and N_2) compete for uploading bandwidth of a source peer (N_S) [Ma et al., 2004a].

where:

$$\sum_{i=1}^{N} x_i \leq W_S \tag{5.9}$$

Here, the logarithmic function represents the utility as perceived by each peer i.

The above optimization problem can be solved by a progressive filling algorithm that prioritizes competing peers in descending order of the marginal utility $C_i/(b_i + x_i)$.

Given values of b_i and C_i, the source peer can compute the allocations in a deterministic manner. However, from the perspective of a requesting peer, a problem remains as to how it should set its bidding value b_i. Using a game theoretic analysis, it is shown that the action profile in which:

$$b_i = \frac{W_S C_i}{\sum_{j=1}^{N} C_j} \forall i$$

is a Nash equilibrium. Furthermore, provided that all cooperative peers use their respective strategies as specified in the Nash equilibrium action profile, collusion among non-cooperative peers can be eliminated. Notice that each requesting peer i needs to know the values of W_S and $\sum_{j=1}^{N} C_j$ in order to determine its own bid b_i. In a practical situation, these two values can be supplied by the source peer to every requesting peer.

5.2.1.8 Auction-Based Approaches

Gupta and Somani [Gupta and Somani, 2004] proposed an auction-based pricing mechanism for P2P file object lookup services. In their model, each resource (e.g., a file object) is stored in a single node. However, the indices for such a file object are replicated at multiple nodes in the network and these nodes are called *terminal nodes*. When a peer initiates a lookup request for a certain file object, the request is sent through multiple paths toward the terminal nodes, as shown in Figure 5.6. The problem here is that the intermediate nodes need some incentives in order to participate in the request forwarding process.

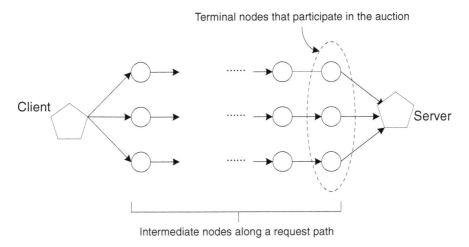

FIGURE 5.6: The request forwarding process [Gupta and Somani, 2004].

Gupta and Somani [Gupta and Somani, 2004] suggested a novel solution to the incentive problem. Specifically, the initiating peer attaches a price in the request message it sends to the first layer of nodes in the request chains. Each intermediate node on the request chains then updates the price by adding its own "forwarding cost." The terminal nodes also do the same updating before sending the request messages to the data source. Upon receiving all the request messages, the data source then performs a second price sealed bid auction (also referred to as Vickrey auction) [Osborne, 2004] to select the highest bid among the terminal nodes. The selected terminal node then needs to pay the price equal to the value of the second highest bid. With this auction-based approach, all the intermediate nodes on the request chains have the incentive to participate in the forwarding process because they might eventually get paid by the requester should their respective request chain wins the auction.

For example, consider the lookup process shown in Figure 5.7. We can see

that the request chain terminated by node T1 wins the auction process and
the payoff to the data source node B is 60. The only intermediate node (node
1) then also gets a payoff. Gupta and Somani [Gupta and Somani, 2004] also
showed that a truthful valuation is the optimal strategy for each intermediate
node. Furthermore, based on the requirement that every message cannot be
repudiated, it is also shown that the proposed mechanism can handle various
potential threats such as malicious auctioneer, collusion between data source
and a terminal node, forwarding of bogus request message, etc.

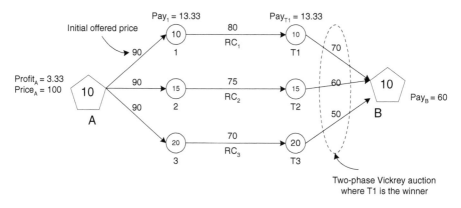

FIGURE 5.7: An example of the auction process in request forwarding
[Gupta and Somani, 2004].

Wongrujira and Seneviratne [Wongrujira and Seneviratne, 2005] also proposed a similar auction-based charging scheme for forwarding nodes on a path
from a requesting peer to a data source. However, they pointed out an important observation that some potentially malicious peers could try to reduce
the profits of other truthful peers by dropping the price messages. To mitigate
this problem, a reputation system is introduced in that every peer maintains
a history of interactions with other peers. The reputation value of a peer is increased every time a message is forwarded by such a peer. On the other hand,
if an expected message exhibits a timeout, the responsible peer's reputation
value is decreased.

Wang and Li [Wang and Li, 2005] also considered a similar problem in
which a peer needs to decide how much to charge for forwarding data. Instead
of using auction, a comprehensive utility function is used. The utility function captures many realistic factors: the quantitative benefits of forwarding
data, the loss in delivering such data, the cost and the benefit to the whole
community. With this utility function, an upstream peer has the incentive to
contribute its forwarding bandwidth while a downstream peer is guided toward
spending the upstream bandwidth economically. Furthermore, a reinforcement
learning component is incorporated so that each peer can dynamically adjust

the parameters in its utility function so as to optimally respond to the current market situations.

Sanghavi and Hajek [Sanghavi and Hajek, 2005] observed that in a typical auction-based pricing mechanism as described above, there is a heavy communication burden on the peers. Indeed, the entire set of user preferences has to be communicated from a peer to the auctioneer. Sanghavi and Hajek then analytically derived a class of alternative information mechanisms that can significantly reduce the communication overhead. Specifically, each peer's bid is only a single real number in each case, instead of an entire real-valued function.

Hausheer and Stiller [Hausheer and Stiller, 2005] studied a completely decentralized auction approach for electronic P2P pricing of goods in a system called PeerMart (which is built on top of Pastry [Rowstron and Druschel, 2001b]). The key idea is the usage of a broker set which comprises other peers in the electronic marketplace. Specifically, a broker set consists of peers whose IDs are closest to the ID of the good in the auction. Each of these peers then potentially acts as the auctioneer in the selling process. The advantage of the broker set-based method is that in case a particular peer in the set is faulty (or even malicious in the sense that it does not respond to auction requests), another member in the set can take up the role of auctioneer. An example is shown in Figure 5.8.

5.2.1.9 Exchange-Based Systems

Motivated by the fact that any payment/credit-based system entails a significant transaction and accounting overhead, Anagnostakis and Greenwald [Anagnostakis and Greenwald, 2004] proposed an exchange-based P2P file sharing system. The fundamental premise is that any peer gives priority to exchange transfers. That is, in simple terms, any peer is willing to send a file to a peer that is able to return a desired file. However, based on this idea, it is incorrect to consider two-way exchanges only. Indeed, a "ring" of exchange involving two or more peers, as shown in Figure 5.9, is also a proper P2P file transfer.

In the exchange-based P2P file sharing system, each peer maintains a data structure called *incoming request queue* (IRQ). Now, a crucial problem is how each peer can determine whether an incoming request should be entertained, i.e., whether such a request comes from some peer on a ring of exchange requests. It is obviously computationally formidable to determine all the potential multi-peer cycles. Fortunately, Anagnostakis and Greenwald [Anagnostakis and Greenwald, 2004] argue that based on simulation results, in practice a peer only needs to check for cycles with up to five peers.

Each peer uses a data structure called *request tree* to check for potential request-cycles. For example, as we can see in Figure 5.10, a peer A decides to entertain a request for file object o_2 because A finds that peer P_9 possesses an object that is needed by A. Based on this checking mechanism, the in-

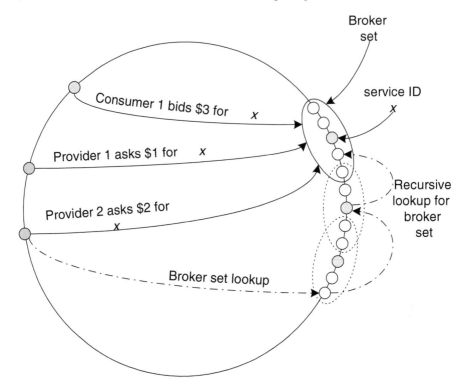

FIGURE 5.8: An example of a fully decentralized auction [Hausheer and Stiller, 2005].

coming requests are prioritized. Simulation results indicate that the proposed exchange-based mechanisms are effective in terms of file object download time.

Table 5.2 gives a qualitative comparison of different incentive approaches proposed for P2P file sharing systems. In general, systems that involve payment would be more difficult to implement because it is not trivial to design a global "currency" for use in such systems. Furthermore, security requirement would be high because the payment could be forged by malicious peers. On the other hand, exchange-based or reciprocity-based are easier to implement and hence, are more scalable. The major crux is that there is much less *state information* to be kept by each peer. More importantly, the accuracy of such state information (e.g., reputation) does not need to be absolutely very high. Thus, we expect that future P2P file sharing would still be based on similar approaches.

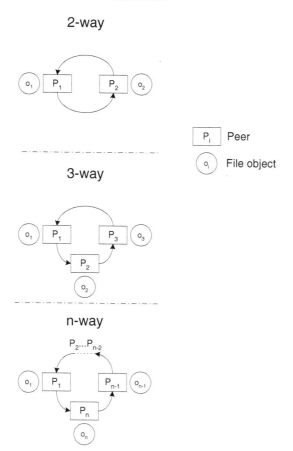

FIGURE 5.9: Different feasible forms of exchanges [Anagnostakis and Greenwald, 2004].

5.2.2   Media Streaming Systems

In this section, we describe several interesting techniques for providing incentives in a P2P media streaming environment. Broadly speaking, there are two different structures employed in P2P media streaming: asynchronous layered streaming and synchronous multicast streaming.

5.2.2.1   Layered Many-to-One Streaming

Xu *et al.* [Xu et al., 2002] proposed a fully distributed differentiated admission control protocol called DAC_{p2p}. In this model, each requesting peer needs multiple supplying peers to send different layers of media data. That is, from a topological perspective, each streaming session involves one single recipient and multiple sources, structured as a two-level inverted tree network.

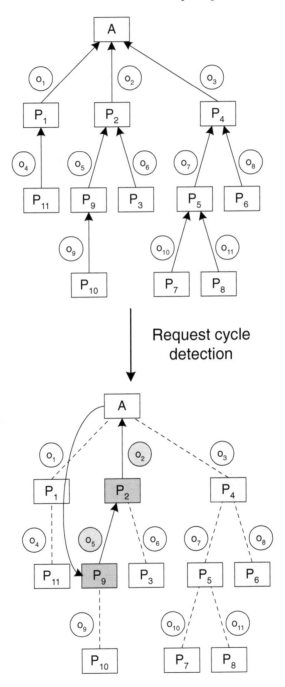

FIGURE 5.10: Request cycle detection using the request tree data structure maintained at each peer [Anagnostakis and Greenwald, 2004].

TABLE 5.2: A qualitative comparison of different incentive approaches for P2P file sharing.

	Payment	Auction	Exchange	Reciprocity	Reputation
Example	Hauscheer et al. [Hausheer et al., 2003]	Gupta and Somani [Gupta and Somani, 2004]	Anagnostakis and Green-wald [Anag-nostakis and Greenwald, 2004]	BitTorrent	Sun and Garcia-Molina [Sun and Garcia-Molina, 2004]
Practical Implementation	Complex	Complex	Easy	Easy	Easy
Security Requirement	High	High	Low	Low	Medium
Centralized Authority Required	Yes	Yes	No	No	Yes/No
Scalability	Medium	Low	High	High	High

Of course, each supplying peer may also be a recipient of another logically different streaming session. With this streaming model, each requesting peer needs to actively contact several supplying peers in order to start a session.

This streaming model is based on a practical observation that peer requests are asynchronous and each peer's communication capability is different (i.e., heterogeneous streaming capabilities), as illustrated in Figure 5.11 where peers with different communication supports (e.g., DSL, dial-up, etc.) initiate streaming requests at different times.

A layer-encoded media streaming process is assumed, as shown in Figure 5.12. As can be seen, each peer performs buffered playback so that the received media data are kept in a buffer and thus, can be used for streaming to other later-coming peers. For example, H_1 initiates a streaming session first and thus, it is served solely by the server. Peer H_2 starts its session next and so it can request H_1, which has buffered some media data, together with the server to send it the required data. Similarly, H_3 can stream from H_1 and H_2 without the server as it starts its session just-in-time to use the buffered data from the two earlier peers. On the other hand, H_4, which starts too late, cannot stream from H_1 and H_2. Instead, it receives media data from H_3 and the server.

Each potential supplying peer has only a limited capacity and thus, an admission control mechanism is needed. Peers in the system are classified into N classes according to the different levels of uploading bandwidth available at the peers. Each potential supplying peer P_S maintains an admission probability vector: $(Pr[1], Pr[2], \ldots, Pr[N])$. Here, a smaller index represents a class with a larger uploading bandwidth. Suppose P_S is itself a class-k peer. Then, its probability vector is initialized as follows:

- For $1 \leq i \leq k$, $Pr[i] = 1.0$;

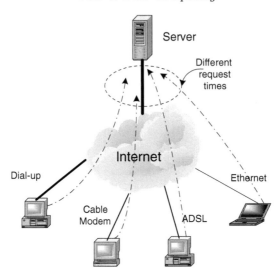

FIGURE 5.11: Asynchrony and heterogeneity of media streaming peers [Cui and Nahrstedt, 2003].

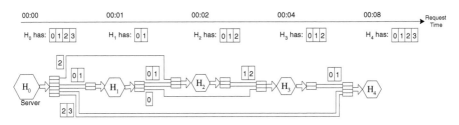

FIGURE 5.12: Layered streaming with buffering for serving asynchronous requests [Cui and Nahrstedt, 2003].

- For $k < i \leq N$, $Pr[i] = \frac{1}{2^{i-k}}$.

Thus, P_S always grants media streaming requests from a higher class peer (i.e., one that has a larger uploading bandwidth). Notice that this is similar in spirit to the incentive approach used in BitTorrent [Cohen, 2003]. For requests from lower class peers, P_S may serve them as governed by the respective probabilities $Pr[i]$ in the vector. If P_S has not served any request during a certain period of time, then the admission probabilities of lower class peers will be increased.

Similar to the case in BitTorrent, each peer has the incentive to report a higher uploading bandwidth because doing so will increase its probability of admission when it needs to initiate a media streaming session. Xue *et al.* [Xue et al., 2004] extended the DAC_{p2p} to a wireless environment. The key idea in the extension is to exploit the spatial distribution of mobile devices to form clusters. Users in a cluster interact using the DAC mechanism.

Habib and Chuang [Habib and Chuang, 2006] also explored a similar idea in providing differentiated peer selection to participating peers. Specifically, a peer has only a limited set of choices (with possibly low media quality) if it behaves selfishly in the system. The degree of selfishness is reflected by a score known to other peers. The score is increased if the peer contributes to other peers, and is decreased if it refuses the requests of other peers. Based on a practical emulation study using the PROMISE [Hefeeda et al., 2003] streaming system implemented on top of PlanetLab [PlanetLab, 2006], Habib and Chuang [Habib and Chuang, 2006] found that the proposed incentive scheme is effective in enhancing the performance of the system.

5.2.2.2 Multicast One-to-Many Streaming

Ngan *et al.* considered an application level multicast system for video streaming. The system is based on SplitStream [Castro et al., 2003a] which in turn is built on top of Pastry [Rowstron and Druschel, 2001b]. The multicast system considered critically relies on a payment-based scheme. Specifically, there are five components:

- **Debt Maintenance:** When a peer A forwards video streaming data to a downstream peer B, B owes A a unit of debt.

- **Periodic Tree Reconstruction:** The multicast tree is reconstructed periodically in order to avoid prolonged unfair connections among peers. An unfair connection is one between a well-behaved peer and a selfish peer.

- **Parental Availability:** Any new peer can obtain location and addressing information about any potential parent peers in the multicast tree. Thus, the new peer can identify a potential selfish parent if the latter consistently refuses connection.

- **Reciprocal Requests:** In the system, any two well-behaved peers are expected to have an equal chance of being parent or child in any given multicast tree.

- **Ancestor Rating:** This is a generalization of the Debt Maintenance component. Here, debts are also accounted for all ancestors of a peer. Specifically, all nodes on a path in forwarding data from the source to a peer are credited or debited in cases where expected data is successfully received or not, respectively.

Simulation results indicate that a selfish free-rider is effectively penalized in terms of the amount of video streaming data received.

Chu and Zhang [hua Chu and Zhang, 2004] also considered a multicast-based streaming environment. The streaming process is synchronous and is supported by a multiple description codec (MDC), in which a server provides several different stripes of video with different quality. A key feature in this streaming environment is that during each streaming session, multiple multicast trees are used, each of which is for sending different stripes of video. This is illustrated in Figure 5.13. With such a streaming structure, a peer can logically join different trees simultaneously at a *different position* in each tree. Specifically, when a peer joins a certain multicast tree at a higher level (e.g., peer A in the tree for stripe II), it needs to provide a larger uploading bandwidth to serve the lower level peers in the same tree. On the other hand, a peer can also join a tree as a leaf so that it becomes a pure recipient in the tree.

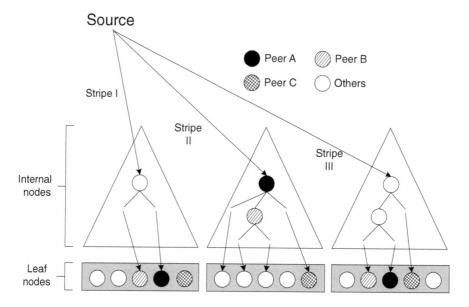

FIGURE 5.13: Layered video streaming using multiple multicast trees [hua Chu and Zhang, 2004].

With this model, Chu and Zhang [hua Chu and Zhang, 2004] then studied the effects of different degrees of altruism. They used a parameter $K = f/r$, where f is the total uploading bandwidth provided by a peer and r is its total downloading bandwidth. Thus, a larger value of K indicates a higher degree of altruism for the peer. Similar to many other P2P sharing systems described above in this chapter, a peer with a higher value of K can enjoy a better performance (in terms of media quality in the streaming application). Simulation results indicate that a small average value of K (e.g., 1.5) can already improve the overall performance of the whole system.

Shrivastava and Banerjee [vivek Shrivastava and Banerjee, 2005] demonstrated that streaming based on a multicast structure could be a result of natural selection. The key idea is depicted in Figure 5.14. The left part of the figure illustrates a situation where multiple peers are sharing the capacity of a single server. As a result, each peer can only enjoy a small downloading data rate. However, when peers are organized as a multicast tree, based on strategic *natural selection* (detailed below), each peer can enjoy a much larger downloading rate.

The natural selection process can be illustrated in Figure 5.15. Here, initially the root is the only source in the system and thus, peer A selects the root as the source, enjoying a downloading rate of 500 kbps. Now, when a new peer B joins the system, peer A has basically two choices: (1) serve peer B; or (2) do not serve peer B. To implement the second choice which seems to be a more favorable one, peer A can declare to the BSE that its uploading rate is 0 or a value smaller than 250Kbps (which is half of the capacity of the root). However, in doing so, peer B has no choice but naturally selects the root to be its streaming source. In that case, peers A and B will share the root's uploading capacity and thus, each obtains only 250Kbps data rate. On the other hand, in anticipation of such an actually unfavorable outcome, peer A should instead strategically declare its uploading bandwidth to be 300Kbps, which is slightly higher than the capacity declared by the root. Consequently, peer A can continue to enjoy a high downloading rate from the root, at the expense of its uploading of data to peer B at a rate of 300Kbps.

Ye and Makedon [Ye and Makedon, 2004] proposed a useful detection and penalty scheme to tackle the existence of selfish peers in a multicast streaming session. They observed that a selfish peer may lie to other peers in that it claims its uploading bandwidth is large so that it can enjoy a higher probability of being admitted into a streaming session or enjoy a higher quality of media data. The key of the detection mechanism is that a downstream peer in a multicast tree returns a "streaming certificate" back to its parent peer. For example, as shown in Figure 5.16, peers P_4, P_5, and P_6, send streaming certificates $\mathbf{SCert}(P_i, P_3)$ to the parent peer P_3 ($i = 4, 5, 6$). The certificates are sent periodically and are time-stamped with authentication. Thus, a higher level peer, e.g., P_1, can periodically check whether its children peers (i.e., P_2 and P_3) are selfish by asking for certificates they have received (if any) from their own children peers. If a peer cannot produce such a certificate, the higher level peer can then remove such a potentially selfish peer from the tree. The removal process is manifested as a termination of media data transmission.

Jun *et al.* [Jun et al., 2005] also explored a similar idea in their proposed Trust-Aware Multicast (TAM) protocol. Targeted for detecting and deterring uncooperative peers which can modify, fabricate, replay, block, and delay data, the TAM protocol is based on a message structure that contains four fields: sequence number, timeout period, data payload, and cryptographic signature. The sequence number is used for detecting duplicated or missing data. The timeout period is used for detecting delayed data. Thus, a selfish (or even malicious) peer can be identified by its children peers in the multicast tree. Different from the approach suggested by Ye and Makedon [Ye and Makedon,

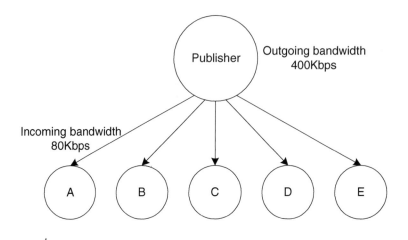

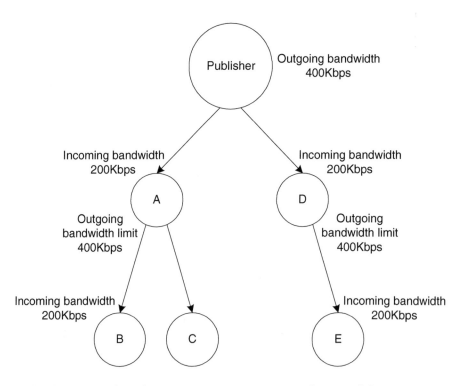

FIGURE 5.14: A multicast streaming structure is better off for every peer [vivek Shrivastava and Banerjee, 2005].

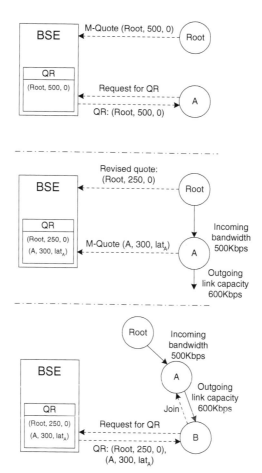

FIGURE 5.15: An illustration of the strategic natural selection process in connecting streaming sources and destinations (BSE is the bootstrap entity providing service information) [vivek Shrivastava and Banerjee, 2005].

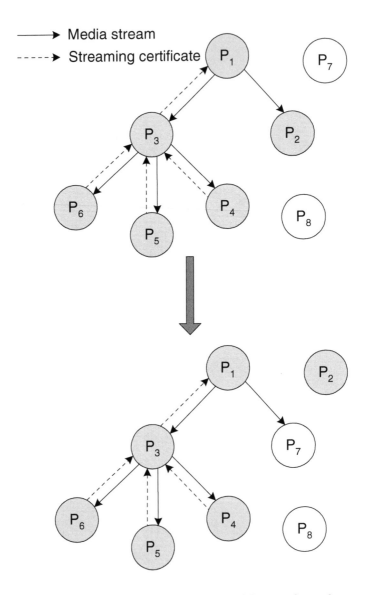

FIGURE 5.16: Detection and removal of a selfish peer from the streaming multicast tree [Ye and Makedon, 2004].

2004], the children peers are responsible for reporting such suspicious selfish peers to the root (or the server) in the tree. Jun *et al.*'s scheme is also more flexible in that even upon receiving such "negative reports," the root may not discard such suspected selfish peers immediately. Instead, the root keeps track of a trust metric for each peer in the tree. A negative report only decreases the trust value. Only when the trust value falls below some threshold, the suspected selfish peer is discarded from the tree and the peers under its subtree are relocated.

5.2.2.3 Coalition-Based Media Streaming

In P2P media streaming, each peer can choose its upstream peers (parents) and downstream peers (children). In a recent study by Yeung and Kwok [Yeung and Kwok, 2009], they consider peers as rational entities and model the peer selection process as a strategic game.

They first focus on the case where there is only one parent, p, and a set of children, $c_1, c_2, \cdots, c_n$. The objective is to study how p should select its children such that the resultant parent-child relationships are stable and resilient to peer dynamics. Specifically, Yeung and Kwok [Yeung and Kwok, 2009] formulate a cooperative game where the players are the parent and its children. The objective is to form a stable coalition which creates the highest aggregate value. Here, stability is defined as the probability that a participant departs from the coalition and acts alone. The aggregate value is to be distributed among the members. In other words, two inter-related issues need to be tackled: (1) formation of a stable coalition; and (2) distribution of the aggregate value. The elements of cooperative game and what constitutes a stable coalition are defined as follows.

A cooperative game consists of a finite set of players, N, and a scalar-valued function, $V(\cdot)$, which associates every subset G of N a real number, $V(G)$. For each coalition, G, the number $V(G)$ represents the total payoff to be divided among the members of G, i.e.,

$$V(G) = \sum_{\forall x \in G} v(x) \tag{5.10}$$

where $v(x)$ represents the value allocated to player x.

Here, $V(G)$ is called the value of the coalition, G. Players can form other coalitions to obtain different values. We say that a stable coalition is formed when players have no incentive to deviate from joining the coalition. Specifically, a coalition, G, is stable if we cannot find a better coalition, G', $G' \subseteq G$, with respect to $V(\cdot)$. This implies:

$$\sum_{x \in G} v(x) \geq V(G') \quad \forall G' \subseteq G \tag{5.11}$$

In other words, if a coalition is unstable, it is possible for a subset of players to deviate such that each deviating player can obtain a larger value than they do staying put. In game theory literature, the above definition of stability is called the *core* of the cooperative game. In our context, it is undesirable for peers to deviate after joining as that would disrupt the structure of the P2P network and, in turn, adversely affect the streaming quality.

With the above definitions, a cooperative game, called the *peer selection game*, can be devised to model the peer selection process. The players are a parent p and a set of children, $c_1, c_2, \cdots, c_n$. The set of all players are denoted as G_a, i.e.,

$$G_a = \{p, c_1, c_2, \cdots, c_n\} \tag{5.12}$$

The players can freely form other coalitions, G, among themselves, where $G \subseteq G_a$. In general, different coalitions lead to different values. The function $V(G)$ should satisfy the following conditions:

$$V(G) = 0 \quad \text{if } p \notin G \tag{5.13}$$

$$V(G) \leq V(G') \quad \text{if } G \subseteq G' \tag{5.14}$$

$$V(G_1 \cup c_i) - V(G_1) \neq V(G_2 \cup c_i) - V(G_2) \tag{5.15}$$

Condition (5.13) dictates that the parent, p, is a necessary member in any coalitions that generate non-zero values. In other words, p is the veto player of the game. This is a reflection of the reality where downstream peers depend on their parent for media packets. Without the participation of p, a coalition does not bring any value to the members.

Condition (5.14) indicates that when comparing two coalitions, G and G', the coalition with more members always generates a value no smaller than the other does. This property precisely models a practical scenario where a parent having a larger number of children is more important because if such a parent departs, a large number of other peers will be disconnected. Thus, the system should attach a higher value to such a coalition.

Condition (5.15) means that, in general, the same peer, c_i, brings different marginal utilities to different coalitions. The discrepancy is attributed to the heterogeneous nature of P2P media streaming. For instance, the presence of c_i would be more significant if the coalition contains only a few children. On the other hand, c_i does not create much value if it joins another coalition already having many children.

We require the value function $V(G)$ to satisfy all the three conditions discussed above. However, the precise definitions depend on the specific characteristics of the application. A specific value function is defined below.

In the peer selection game, the formation of a coalition G, would create an aggregate value, represented by $V(G) = \sum_{\forall x \in G} v(x)$, where $v(x)$ represents the utility allocated to player x. It is assumed that each player would like to maximize its share of utility, i.e., player x is interested in maximizing $v(x)$. This is reasonable because each peer x is more concerned with its own performance in terms of $v(x)$. On the other hand, $V(G)$ is a measure of the value of coalition G, in the P2P network.

The participating cost of peers should also be taken into consideration.

Specifically, player x incurs some cost to be a member of a coalition. The amount of player x's coalitional effort is denoted as $e(x)$. This can be interpreted as the amount of outgoing bandwidth and other resources consumed. The utility of player x is then defined as the difference between the share of value obtained from the coalition $v(x)$, and the amount of effort contributed to the coalition $e(x)$. That is, utility is defined as:

$$u(x) = v(x) - e(x) \qquad (5.16)$$

Moreover, it is assumed that $e(x)$ depends on the number of peers in the coalition, i.e.,

$$e(x) = \begin{cases} (|G| - 1)e & x = p \\ e & x \in G \setminus \{p\} \end{cases} \qquad (5.17)$$

where e is a non-negative constant.

It is clear that if player x does not join any coalition, its utility is zero, i.e., $u(x) = 0$ if $x \notin G$. This implies that a rational player will only join a coalition providing non-negative utility, $u(x) \geq 0$ if $x \in G$. This is called the *incentive compatibility* constraint:

$$u(x) \geq 0 \text{ if } x \in G \qquad (5.18)$$

Here, G_a is defined as the set of players and can be used to analyze the peer selection game as G_a increases:

Case 1 $G_a = \{p\}$

This is the baseline case where the parent is the sole player. There is only one possible coalition, $G_1 = \{p\}$. The player obtains all the value created by the coalition, which is given by:

$$V(G_1) = v(p) \qquad (5.19)$$

Since p has no downstream peer, its effort is zero, i.e., $e(p) = 0$. The utility of p is $u(p) = v(p)$.

Case 2 $G_a = \{p, c_1\}$

The set of players includes the parent and one potential child, i.e., $P = \{p, c_1\}$. If p accepts c_i as its child, they form a coalition, $G_2 = \{p, c_1\}$. The value created by the coalition is to be distributed between the two players, i.e.,

$$V(G_2) = v(p) + v(c_1) \qquad (5.20)$$

The value $V(G_2)$ needs to be distributed judiciously in order to make G_2 a stable coalition such that neither p nor c_1 has an incentive to leave. This requires the following conditions to be satisfied:

$$v(p) - e \geq V(G_1) \qquad (5.21)$$
$$v(c_1) - e \geq 0 \qquad (5.22)$$

Condition (5.21) suggests that p should receive a utility larger than the value created by acting alone. Condition (5.22) requires that the share of value allocated to c_1 should be at least the amount of its contributed effort.

In other words, the share of value allocated c_1, denoted by $v(c_1)$, should be:

$$e \leq v(c_1) \leq V(G_2) - V(G_1) - e \qquad (5.23)$$

<u>Case 3</u> $G_a = \{p, c_1, c_2\}$

The set of players now includes p and two potential children, i.e., $P = \{p, c_1, c_2\}$. If the parent accepts both peers, they form a larger coalition, G_3, and create a value of $V(G_3)$. This is to be distributed among the three players:

$$V(G_3) = v(p) + v(c_1) + v(c_2) \qquad (5.24)$$

It should be ensured that G_3 is a stable coalition where the parent and the two children have no incentive to leave. This requires the following conditions to be satisfied:

$$
\begin{aligned}
v(p) - 2e &\geq V(G_1) & (5.25) \\
v(c_1) - e &\geq 0 & (5.26) \\
v(c_2) - e &\geq 0 & (5.27) \\
v(p) + v(c_1) &\geq V(\{p, c_1\}) & (5.28) \\
v(p) + v(c_2) &\geq V(\{p, c_2\}) & (5.29)
\end{aligned}
$$

Condition (5.25) ensures that the parent would not drop the two children. Conditions (5.26) and (5.27) lead to non-negative utilities for c_1 and c_2, respectively. In other words, these two conditions are the incentive compatibility constraint in (5.18). The last two conditions, on the other hand, cause dropping one of the two children an undesirable move. The conditions can be simplified as follows:

$$
\begin{aligned}
v(c_1) &\leq V(G_3) - V(\{p, c_2\}) & (5.30) \\
v(c_2) &\leq V(G_3) - V(\{p, c_1\}) & (5.31) \\
v(c_1) + v(c_2) &\leq V(G_3) - V(G_1) - 2e & (5.32) \\
v(c_1), v(c_2) &\geq e & (5.33)
\end{aligned}
$$

<u>Case n</u> $G_a = \{p, c_1, \cdots, c_{n-1}\}$

This is the general scenario where the parent is encountered with $(n-1)$ potential children. If they form a single coalition of size n, this creates a value of $V(G_n)$, which is to be distributed among the members, i.e.,

$$V(G_n) = v(p) + \sum_{c_i \in G_n} v(c_i) \qquad (5.34)$$

For G_n to be stable, peers should have no incentive to leave the coalition individually or as a group. Similar to previous cases, the following conditions can be obtained:

$$v(c_r) \leq V(G_n) - V(G_n \setminus \{c_r\}) \quad \forall c_r \tag{5.35}$$

$$\sum_{\forall c_i \in P} v(c_i) \leq V(G_n) - V(G_1) - (n-1)e \tag{5.36}$$

$$v(c_r) \geq e \quad \forall c_r \tag{5.37}$$

The term "$V(G_n) - V(G_n \setminus \{c_r\})$" is called the marginal utility of c_r. It is the additional amount of value created by c_r to the original coalition. Since p's effort is increased by e if c_r is accepted as its child, the share of value allocated to c_r is:

$$v(c_r) = V(G_n) - V(G_n \setminus \{c_r\}) - e \tag{5.38}$$

A Specific Value Function

A specific value function for the peer selection game is proposed [Yeung and Kwok, 2009]:

$$V(G) = \begin{cases} \log(1 + \sum_{\forall i \neq p} \dfrac{1}{b_i}) & p \in G \\ 0 & \text{otherwise} \end{cases} \tag{5.39}$$

Without loss of generality, the value function is zero when the parent is the sole coalition member, i.e., $V(G_1) = 0$. This is an increasing function in coalition size. In other words, a new peer always brings additional value to an existing coalition. Furthermore, a peer may create different values to different coalitions. Therefore, the value function satisfies Conditions (5.13), (5.14), and (5.15).

Besides the above characteristics, the value function can also differentiate peers according to their outgoing bandwidth values. For the same coalition, G, peer x would receive a larger share of the value than peer y if $b_x < b_y$. The reason for that arrangement would become evident with the following numerical example.

Consider two coalitions: G_X and G_Y where $G_X = \{p_x, c_1, c_2\}$ and $G_Y = \{p_y, c_3, c_4, c_5\}$. A peer c_6 would like to join one of the two coalitions. We take $e = 0.01$ and the outgoing bandwidths of the peers are listed as follows:

b_1	b_2	b_3	b_4	b_5	b_6
1	2	2	2	3	2

It is easy to see that $V(G_X) = 0.92$ and $V(G_Y) = 0.85$. If c_6 joins the coalition G_X, we have $V(G'_X) = 1.10$ and its share of value is: $V(G'_X) - V(G_X) - e = 0.17$. On the other hand, c_6 joining coalition G_Y would result in $V(G'_Y) = 1.04$ and its share of value is: $V(G'_Y) - V(G_Y) - e = 0.18$. Therefore, c_6 joins G_Y and $v(c_6) = 0.18$. The peer's share of value, i.e., $v(x)$, in the coalition is then used by the parent in determining the amount of bandwidth allocation [Yeung and Kwok, 2009].

5.3   Incentive Issues in Wireless P2P Systems

Wireless P2P systems [Hsieh and Sivakumar, 2004] are proliferating in recent years. Thanks to the widely available hot-spot wireless environments, users handheld devices can work with each other in an ad hoc and impromptu manner. As will be evident in this section, many techniques designed for wired environments are also applied in a wireless system in a similar manner. Nevertheless, there is a unique challenge in a wireless environment, namely the connectivity issue. Furthermore, there is also one more dimension of cost incurred in each wireless P2P user—the energy expenditure, which is of prime concern to the user as wireless devices are largely powered by batteries.

5.3.1   Routing and Data Forwarding

In a wireless P2P system, the connectivity among peers is itself a bootstrap sharing problem. Indeed, if wireless users are unwilling to cooperate in performing routing and data forwarding, the wireless network can become partitioned so that service providers cannot be reached by potential service consumers. In view of this critical challenge, there has been a plethora of important research results related to incentive issues for ad hoc routing and data forwarding. In the following, we briefly cover several interesting techniques that are based on payment mechanisms, auction mechanisms, reputation systems, and game theoretic modeling.

Ileri *et al.* [Ileri et al., 2005] proposed a payment-based scheme for enticing devices to cooperate in forwarding data for other devices in the network. The payment is not in monetary terms but in terms of bits-per-Joule. Specifically, the utility of a user i in the network is defined as:

$$u_i(p_i) = \frac{T_i(p_i)}{p_i} \tag{5.40}$$

where u_i is the utility, p_i is the transmit power, and T_i is the throughput. That is, the utility is equal to the average amount of data received per unit energy expended, also in bits-per-Joule.

The payment system also involves an access point in the wireless network. Specifically, the access point also tries to maximize its revenue by using two parameters, μ and λ, judiciously. Here, λ is the unit price of service provided by the access point to any device in the network. On the other hand, μ is the unit reimbursement the access point provides to any device which has helped forward other devices' traffic. The access point's revenue is therefore given by:

$$\rho = \sum_{\text{all users } i} \lambda T_{in_i} - \sum_{\text{all forwarders } j} \mu T_{ja}^{eff-for} \tag{5.41}$$

where T_{in_i} is the service provided by the access point to user i and $T_{ja}^{eff-for}$

is the service forwarded by a forwarder j to the access point. The situation is as shown in Figure 5.17.

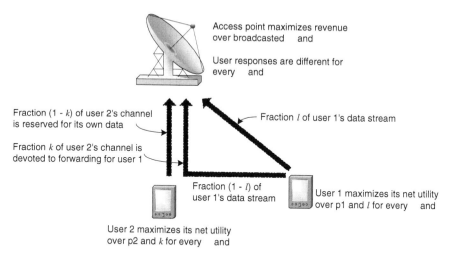

FIGURE 5.17: Charging of network service and reimbursement of data forwarding [Ileri et al., 2005].

Simulation results indicate that the proposed service reimbursement scheme generally improves the network aggregate utility.

Salem *et al.* [Salem et al., 2006] also considered a payment-based scheme for encouraging cooperation. However, their scheme involves real monetary costs and a payment clearance infrastructure (e.g., a billing account for each user). The charging and rewarding scheme is similar to that we described above—a forwarder will get reimbursed and a normal user using network service will get charged.

Marbach and Qiu [Marbach and Qiu, 2005] investigated a similar problem based on individual device pricing. However, the main differences are that each device is allowed to freely decide how much to charge for forwarding traffic (in previous researches, usually the unit price is the same for all devices) and there is no budget constraint on all the devices.

Wang and Li [Wang and Li, 2005] proposed an auction-based scheme similar to the work by Gupta and Somani [Gupta and Somani, 2004] that we described in Section 5.2.1.8. Specifically, each device in a wireless ad hoc network declares its cost for forwarding data when some other device wants to initiate a multihop transmission. After considering all possible paths that are able to reach the destination device, the least cost path is chosen and the devices on the path are paid for their forwarding. Again using a VCG-based analysis, it is shown that the dominant action for each device is to report its true cost of forwarding. Wang and Li also showed that no truthful mechanism can avoid collusion between two neighboring devices in the forwarding auction game.

Buchegger and Le Boudec [Buchegger and Boudec, 2005] observed that economic incentives such as payment approaches can entice selfish users to help in routing and forwarding data but may not be able to handle other types of misbehaviors such as packet dropping, modification, fabrication, or timing problems. Thus, they proposed to use a reputation system in which every user provides "opinion" data to the network based on observing the behaviors of neighboring devices. After a user device has gathered such opinions (both from itself or from others, i.e., second-hand information), it can carry out a Bayesian estimation so as to classify the neighboring devices as malicious or normal. A neighboring device that is identified as a malicious user is then isolated from the network by rejecting its routing and forwarding requests.

Felegyhazi *et al.* [Felegyhazi et al., 2006] reported an interesting game theoretic analysis of the forwarding problem in ad hoc networks. Instead of using a payment-based strategy, the model employs a purely utility concept in that a device's utility is equal to its payoff when it acts as a data source (i.e., the sender of a multihop traffic), minus the cost when it acts as an intermediate device (i.e., a forwarder of other sender's traffic) in any time-slot. Here, both the payoff and cost are defined in terms of data throughput. Thus, an important assumption in this model is that only the sender has a positive payoff, while all the intermediate devices enjoy no payoff but just incur forwarding costs. Specifically, the destination device (i.e., the receiver of the multihop traffic) also enjoys no payoff. This may not conform to a realistic situation. Simulations were done to estimate the probability that the conditions for a cooperative equilibrium hold in randomly generated network scenarios.

Table 5.3 gives a qualitative comparison of various data forwarding approaches in wireless ad hoc networks. In general, some form of payment is required. However, as the devices in a wireless ad hoc network are not under a centralized authority's control, it is very difficult to enforce a secure payment clearance mechanism. Auction schemes are interesting but are also difficult to implement in practice because a highly trusted communication infrastructure is required for exchanging bidding information. Yet this is a paradoxical requirement as the communication among wireless peers is itself the ultimate goal in data forwarding. Similarly, a reputation-based approach, while not difficult to implement in practice, could also lead to a paradoxical situation in the sense that the reputation values may not be trustworthy.

5.3.2   Wireless Information Sharing Systems

Wolfson *et al.* [Wolfson et al., 2004] investigated an interesting opportunistic wireless information exchange problem in which a moving vehicle transmits the information it has collected to encountered vehicles, thereby obtaining other information from those vehicles in exchange. The incentive mechanisms

TABLE 5.3: A qualitative comparison of various wireless ad hoc data forwarding approaches.

Proposed Approach	Incentive Scheme	Implementation Difficulty	Security Required	Major Drawback
Ileri *et al.* [Ileri et al., 2005]	Payment	Low	High	Energy-Based Currency
Salem *et al.* [Salem et al., 2006]	Payment	High	High	Real Money
Marbach and Qiu [Marbach and Qiu, 2005]	Payment	High	High	Unlimited Budget
Wang and Li [Wang and Li, 2005]	Auction	High	High	Communication
Buchegger and Le Boudec [Buchegger and Boudec, 2005]	Reputation	Low	Low	Trust
Felegyhazi *et al.* [Felegyhazi et al., 2006]	Utility	Low	Low	Receiver's Payoff

suggested are based on virtual currency. Specifically, each mobile user carries some virtual currency in the form of a protected counter. Two different mechanisms are considered: producer-paid and consumer-paid. In the former, the producer of information pays while the consumer pays in the latter. The price P of a piece of information item (e.g., availability information about a parking lot) is given by:

$$P = E - t - \frac{d}{v} \tag{5.42}$$

where E is the gross valuation of the information item, t is the time elapsed since the information item is created, d is the distance to travel before the user of the information can reach the relevant location (e.g., the parking space), and v is the speed of the user. Simulation results under a simple situation where there is only one consumer and two parking spaces indicate that the proposed incentive mechanisms are effective.

Yeung and Kwok [Yeung and Kwok, 2006a] considered an interesting scenario in wireless data access: a number of mobile clients are interested in a set of data items kept at a common server. Each client independently sends requests to inform the server of its desired data items and the server replies in the broadcast channel. Yeung and Kwok investigated the energy consumption characteristics in such a scenario.

Figure 5.18 depicts the system model for wireless data access. It consists of a server and a set of clients, N. The clients are interested in a common set of data items, D, which are kept at the server. To request a specific data item, d_a, client i is required to inform the server by sending an uplink request, represented by $q_i(d_a)$. The server then replies with the content of the requested data item, d_a, in the common broadcast channel. This allows the data item to be shared among different clients. As illustrated in Figure 5.18, both clients

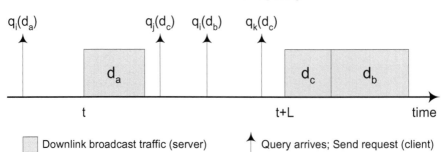

FIGURE 5.18: System model for wireless data access [Yeung and Kwok, 2006a].

j and k request the same data item, d_c, in the second interval. However, the server is required to broadcast the content of d_c only *once* in the next broadcast period, which reduces the bandwidth requirement.

To successfully complete a query, a client expends its energy in two different parts: (1) informing the server of the desired data item, E_{UL}; and (2) downloading the content of the data item from the common broadcast channel, E_{DL}. The energy cost of sending a request to the server is represented by E_s. If a client sends a request for each query, then E_{UL} would be the same as E_s. However, we show that clients need not send requests for each query even without caching. This implies that E_{UL} does not necessarily equal E_s. In general, the total energy required to complete a query, E_Q, is given by:

$$E_Q = E_{UL} + E_{DL} \qquad (5.43)$$

It is assumed that E_{DL} is the same for all clients, but its value depends on the size of a data item. In practice, E_s is a function of various quantities, including spatial separation, speed, instantaneous channel quality, bit-error rate requirement, etc. For simplicity, however, E_s is also assumed to be a fixed quantity.

Based on this model of wireless data access, a novel utility function for quantifying performance is defined as follows:

$$U_i = \frac{E_{total}}{E_Q^i} \qquad (5.44)$$

where E_{total} is the total energy available. The objective is then to reduce the amount of energy consumed in the query process such that *every* client's utility (Equation (5.44)) is increased.

Based on the utility function, Yeung and Kwok formulate the wireless data access scheme as a non-cooperative game—*wireless data access (WDA) game*. Game theoretic analysis shows that while the proposed scheme does not rely on client caching, clients do not always send requests to the server. Simulation results also suggest that the proposed scheme, compared with a simple always-request one, increases the utility and lifetime of *every client* while reducing the number of requests sent, at the cost of slightly larger average query delay. They also compared the performance of the proposed scheme with two popular schemes that employ client caching. The simulation results show that caching only benefits clients with high query rates while resulting in both shorter lifetime and smaller utility in other clients.

5.3.3 Network Access Sharing

Efstathiou and Polyzos [Efstathiou and Polyzos, 2003] studied the problem of building a federation of wireless networks using a fully autonomous P2P approach. Specifically, in their system model, there are multiple WLANs, each of which is considered to be completely autonomous. When a user of one WLAN enters the domain of a different nearby WLAN, the latter would also admit the user based on a reciprocity idea. To achieve this, each WLAN is equipped with a domain agent (DA) which is responsible for managing the roaming of foreign users. Each DA maintains a counter of tokens, which is increased when a foreign user is admitted to its WLAN, and is decreased when a local user travels to a foreign WLAN. The DAs of different WLANs interact with each other in a pure P2P fashion. Thus, the advantage of this approach is that there is no need to set up prior *pairwise* administrative agreements among different WLANs.

Based on a prototype built using Cisco WLANs, it is found that the proposed P2P-based roaming scheme is efficient.

Kang and Mutka [Kang and Mutka, 2005] considered an interesting problem in which peers share the access cost of wireless multimedia contents. Specifically, one peer in the network serves as a proxy which pays a network server in order to download some multimedia contents in a wireless fashion. Other peers in the system then share the contents without incurring any cost. This is achieved by having the proxy broadcast the received multimedia data to the peers within its transmission range. For other distant peers, rebroadcasting by the edge peers is employed to serve them. This is illustrated in Figure 5.19 (left side).

This idea of cost sharing in wireless data access is called CHUM (cooperating ad hoc networking to support messaging). A key component in such sharing is the cost sharing mechanism. In the CHUM system, the peers take turn, in a round-robin manner, to serve as the proxy. This is illustrated in Figure 5.19 (right side). It is assumed that the peers have the incentive to follow this round-robin rule based on the reciprocity concept. Simulation results indicate that 80% of network access cost is saved even with just six peers in the system.

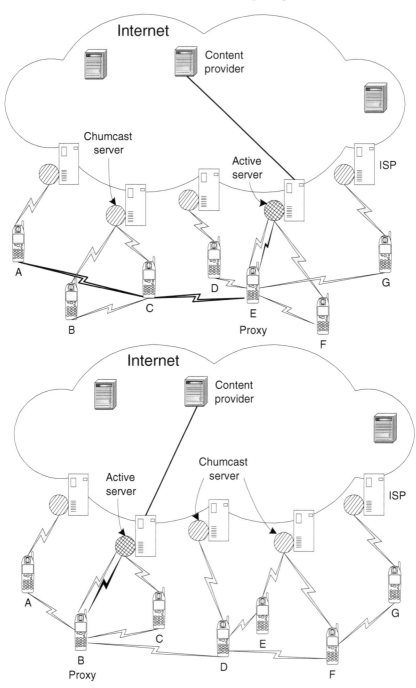

FIGURE 5.19: Illustration of network access cost sharing (left), and round-robin scheduling of proxy (right) [Kang and Mutka, 2005].

5.3.4   Wireless P2P Media Streaming

Yeung and Kwok [Yeung and Kwok, 2008] studied an interesting incentive protocol for energy efficient wireless P2P media streaming. In a wireless media streaming application, users obtain media feeds from subscribed servers for video entertainment or useful information access (e.g., video clips for important stock market information). However, such convenience inevitably comes with a price in that the energy consumption of the device is greatly increased due to the continuous isochronous nature of media streaming and the high volume of data involved (e.g., in video streaming). Thus, energy efficiency is of a prime concern in supporting media streaming applications.

For this problem of providing energy efficient media streaming to mobile users, an insight is that we can consider the availability of multiple wireless networking interfaces in devices nowadays. For instance, it is now quite common for a commodity wireless gadget to have at least two wireless interfaces, e.g., a CDMA2000 cellular interface and an IEEE 802.11x wireless LAN (WLAN) interface. Here, we call the former kind of interface the *server interface* while the latter the *client interface*. A key observation is that the energy consumption characteristics of the two interfaces are very different [Cisco, 2009, GTRAN, 2009]. Specifically, the peer interface consumes less energy to deliver the same amount of traffic than the server interface does. Notice that some higher end devices can even have more than one client interface, e.g., having both WLAN and Bluetooth.

Equipped with two heterogeneous networking interfaces, devices can form a hybrid wireless network in which some devices are connected to wireless servers (via cellular base-stations) using the server interfaces while connecting to other devices using the client interfaces. Some devices may only connect to other wireless peers but not connect to the servers. While this hybrid wireless networking infrastructure is feasible and interesting, how it can help in supporting energy efficient media streaming is still a largely unexplored research issue.

5.3.4.1   System Model

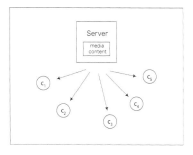

FIGURE 5.20: System model—a media server and a set of mobile clients.

Let us consider a non-interactive media streaming application in a hybrid wireless network, which consists of a server and a set of mobile clients, as shown in Figure 5.20. In this scenario, the primary QoS requirement is to provide clients with sufficient bandwidth in order to achieve uninterrupted streaming. The effect of delay jitter could be largely compensated by having enough playback buffers. Thus, we can focus on the amount of bandwidth from the server to clients.

Each mobile client uses its server interface to access the server, while the peer interface allows neighboring clients to communicate with one another. The clients are interested in a piece of media content owned by the server. We represent the rate of the media content as $r$kbps and its duration as ts. Here, the server splits the media content into n stripes using a multiple description coding scheme. Each stripe is then delivered as a separate stream of media packets. These n stripes are assumed to be independent and of equal rate. This arrangement allows heterogeneous clients to adjust their streaming quality by subscribing to a different number of stripes. The primary QoS requirement is to maintain sufficient bandwidth in order to receive those media packets from the subscribed stripes. Besides satisfying the bandwidth requirement, we focus on the energy cost of clients in obtaining the media packets. Specifically, we treat the QoS requirement as a constraint for optimizing the energy consumption in the clients.

Ignoring control overheads, the energy cost of receiving the complete media content, i.e., all n stripes, through the server interface is (Table 5.6 summarizes the list of symbols and their definitions):

$$E_s^{RX} = \frac{P_s^{RX}}{R_s} \times r \times t \tag{5.45}$$

Correspondingly, the energy cost of receiving the complete media content through the peer interface is:

$$E_p^{RX} = \frac{P_p^{RX}}{R_p} \times r \times t \tag{5.46}$$

Tables 5.4 and 5.5 show the technical specifications of typical server and peer interfaces, respectively. Using these numbers, we have: $E_s^{RX} = 103.13$J and $E_p^{RX} = 8.62$J. This suggests that it is possible to utilize the peer interface for energy efficient media streaming in a hybrid wireless network. In particular, it is interesting to study how heterogeneous clients collaborate to stream the media content from the server.

To quantify heterogeneity, we represent the *type* of client x as α_x, which is defined as:

$$\alpha_x = \frac{\text{amount of energy client } x \text{ willing to consume}}{E_s^{RX}(n)} \tag{5.47}$$

Depending on client x's preferences, α_x may be any non-negative value.

TABLE 5.4: Technical specifications of a typical server interface.

GTRAN DotSurfer 6210 [GTRAN, 2009] (1xEV-DO Release 0)	
Voltage	3.3V
Receive current	150mA
Receive power	495mW
Data rate	2.4Mbps

TABLE 5.5: Technical specifications of a typical peer interface.

Cisco AIR-CB21AG [Cisco, 2009] (IEEE 802.11a/b/g)	
Voltage	3.3V
Transmit current	530mA
Transmit power	1749mW
Receive current	282mA
Receive power	930.6mW
Data rate	54Mbps

For example, $\alpha_x > 1$ means that x is willing to consume more energy than the cost of subscribing to all n stripes ($E_s^{RX}(n)$), e.g., a mobile device equipped with plentiful energy resources. On the other hand, $\alpha_x \leq 1$ means that x is more concerned with the energy cost than the streaming quality, e.g., a mobile device with little residual energy. Specifically, the value of α_x determines the number of stripes client x would subscribe to. Since the n stripes are assumed to be of equal rate, i.e., r/nkbps, the energy cost of streaming i stripes via the server interface is:

$$E_s^{RX}(i) = \frac{i}{n} \times \frac{P_s^{RX}}{R_s} \times r \times t \qquad (5.48)$$

This implies that client x would stream up to i stripes from the server if $\alpha_x \geq \alpha_1(i)$, where $\alpha_1(i)$ is given by:

$$\alpha_1(i) = \frac{E_s^{RX}(i)}{E_s^{RX}} = \frac{i}{n} \qquad (5.49)$$

The set of $\alpha_1(i)$, $i \in [0, n]$, are the threshold values when client x independently streams from the server. Figure 5.21 shows the variation of the number of subscribed stripes with the type of client for $n = 10$. For example, if $\alpha_x = 0.55$, client x would subscribe to 5 stripes from the server. This represents the performance of media streaming when each client acts independently. However, neighboring clients could utilize their peer interfaces to improve streaming performance without violating their energy consumption

TABLE 5.6: Symbols.

Symbol	Definition
r	rate of the media content
t	duration of the media content
n	number of stripes
P_s^{RX}	receive power of the server interface
R_s	data rate of the server interface (downlink)
P_p^{RX}	receive power of the peer interface
R_p	data rate of the peer interface (symmetric)
$E_s^{RX}(i)$	energy cost of receiving i stripes through the server interface ($i \leq n$)
$E_p^{RX}(i)$	energy cost of receiving i stripes through the peer interface ($i \leq n$)
$E_p^{TX}(i)$	energy cost of transmitting i stripes through the peer interface ($i \leq n$)
α_x	client x's type
$\alpha_1(i)$	threshold value for a client to receive i stripes from the server
$\alpha_M(i)$	threshold value for a master to receive i stripes, $i \leq n$
$\alpha_S(i)$	threshold value for a slave to receive i stripes, $i \leq n$
$\alpha_r'(i)$	threshold value for a coordinator with $(r-1)$ helpers to receive i stripes
$\alpha_r(i)$	threshold value for a helper to receive i stripes

constraints. Since clients are autonomous entities, we should also consider their incentives for collaboration. Specifically, each client is modeled as a selfish but rational entity whose degree of selfishness is characterized by its type. Here, client x would only collaborate with other clients provided that the number of stripes x obtained from collaboration is no smaller than s_1, where s_1 is given by:

$$s_1 = \mathrm{argmax}_{\forall i \leq n}\{\alpha \leq \alpha_1(i)\} \qquad (5.50)$$

This means that each client is only interested in improving its own streaming performance. We note that the type of a client quantifies its selfishness. For example, when $\alpha_x = 0.8$, client x would collaborate with other clients for media streaming if it can receive more stripes from collaboration; otherwise, x would rather stream the media content from the server independently. It is then interesting to study the feasibility of collaboration among these selfish clients.

As in all practical P2P data sharing systems (e.g., BitTorrent), there are always some enthusiastic participants that would be willing to spend more resources in return of a larger participating population. Such enthusiastic clients can be modeled as *masters* which have a larger value of α, as detailed below.

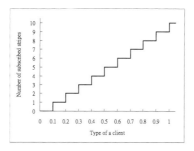

FIGURE 5.21: Number of subscribed stripes versus the type of a client $(n = 10)$.

5.3.4.2 Two Neighboring Clients

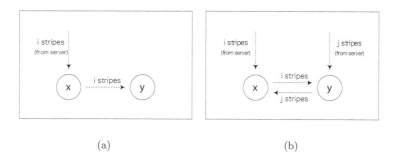

(a)   (b)

FIGURE 5.22: A scenario with two neighboring clients. (a) Master-slave. (b) Peer-to-peer.

First, let us consider a simple scenario with two neighboring clients: $\{x, y\}$. There are two cases: (1) x and y form a master-slave relationship, as shown in Figure 5.22(a); (2) x and y form a peer-to-peer relationship, as shown in Figure 5.22(b).

Without loss of generality, let us assume that x is the master while y is the slave. As depicted in Figure 5.22(a), x subscribes to i stripes from the server and sends them to y through the peer interface. In this case, x is an enthusiastic client, serving y at the expense of its own energy resources. This requires that the type of master x should satisfy: $\alpha_x \geq \alpha_M(i)$, where $\alpha_M(i)$ is given by:

$$\alpha_M(i) = \frac{E_s^{RX}(i) + E_p^{TX}(i)}{E_s^{RX}} \qquad (5.51)$$

On the other hand, client y may decide to "free-ride" because its residual energy is low. This requires that the type of slave y should satisfy: $\alpha_y \geq \alpha_S(i)$, where $\alpha_S(i)$ is given by:

$$\alpha_S(i) = \frac{E_p^{RX}(i)}{E_s^{RX}} \tag{5.52}$$

From Equations (5.51) and (5.52), the threshold values for master and slave as α varies between 0 and 1.2, are shown in Figure 5.23. When two neighboring clients whose types satisfy the two thresholds collaborate, they form a master-slave relationship for media streaming. We can see that the type of a slave is much smaller than that of a master. This is because the slave does not contribute but relies mostly on the master for all media packets. On the other hand, the energy cost of being a master is higher than that of acting alone, which leads to the increase in the threshold values. Let us illustrate the master-slave relationship with the following numerical example.

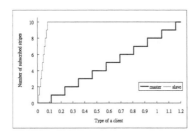

FIGURE 5.23: Master-slave: number of subscribed stripes versus the type of a client ($n = 10$).

Consider two neighboring clients: x and y whose types are 1.2 and 0.25, respectively. If each of them independently streams the media from the server, x will subscribe to 10 stripes while y will subscribe to 2 stripes. However, they may collaborate and form a master-slave relationship for media streaming, as illustrated in Figure 5.22(a). Specifically, x becomes the master and y is the slave. Equation (5.51) suggests that x subscribes to 10 stripes and also sends them to y, which receives the 10 stripes via its peer interface.

The master-slave collaboration arrangement allows the slave to take advantage of the generosity of the master, which provides the media content through its peer interface. The performance of the slave is improved at the expense of the master's energy resources. However, it would be more interesting if both clients contribute their resources to form a peer-to-peer relationship, as depicted in Figure 5.22(b). Here, we can assume that client x and y subscribe to i and j stripes from the server, respectively. They periodically exchange their stripes with each other using the peer interfaces. Effectively, each client obtains $(i+j)$ stripes of the media content, where $(i+j) \leq n$. This peer-to-peer collaboration arrangement improves the performance of both clients. However, the values of i and j depend on the type of the corresponding clients. If client x subscribes to i stripes from the server, we require: $\alpha_x \geq \alpha_2(i)$, where $\alpha_2(i)$

is given by:

$$\alpha_2(i) = \frac{E_s^{RX}(i) + E_p^{TX}(i) + E_p^{RX}(j)}{E_s^{RX}} \tag{5.53}$$

Similarly, we require: $\alpha_y \geq \alpha_2(j)$, where $\alpha_2(j)$ is given by:

$$\alpha_2(j) = \frac{E_s^{RX}(j) + E_p^{TX}(j) + E_p^{RX}(i)}{E_s^{RX}} \tag{5.54}$$

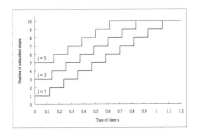

FIGURE 5.24: Peer-to-peer: number of subscribed stripes versus the type of a client ($n = 10$).

Figure 5.24 shows the variation of the number of subscribed stripes with the type of client x, α_x, for $j = 1, 3, 5$. This means that if $\alpha_x = 0.5$, x will obtain 7 stripes when $j = 3$, i.e., $i = 4$. Let us illustrate the peer-to-peer relationship with the following numerical example.

Consider two neighboring clients: x and y whose types are 0.75 and 0.55, respectively. If they independently stream the media from the server, x will subscribe to 7 stripes while y will subscribe to 5 stripes. However, they may collaborate and form a peer-to-peer relationship for media streaming, as shown in Figure 5.22(b). With reference to Equations (5.53) and (5.54), x subscribes to 6 stripes from the server and y subscribes to another 4 stripes from the server. Besides that, they periodically exchange media packets via their peer interfaces. This effectively allows them to receive the complete 10 stripes, i.e., the complete media content.

With the peer-to-peer relationship, both clients increase the number of received stripes without violating their types. Thus, the collaboration between two neighboring clients would improve the performance of media streaming under the same energy consumption constraints.

5.3.4.3 Three Neighboring Clients

Next, let us consider the scenario with three neighboring clients: $\{x, y, z\}$, where y and z are the neighbors of x but they may not be able to communicate with each other directly. Similar to the two-client scenario, the collaboration among the three clients can take on two different forms: master-slave and peer-to-peer.

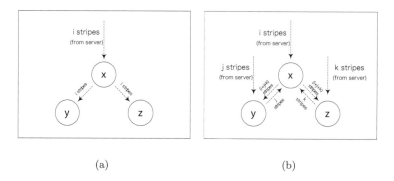

(a)   (b)

FIGURE 5.25: The scenario with three neighboring clients. (a) Master-slave. (b) Peer-to-peer.

Without loss of generality, let us assume that x is the master while y and z are the slaves. Specifically, x subscribes to i stripes from the server and sends them to y and z through the peer interface, as depicted in Figure 5.25(a). Because the master can broadcast the media content to its slaves, the energy cost of being a master does not change with the number of slaves. It follows that the threshold values for both master and slave remain the same, i.e., $\alpha_x \geq \alpha_M(i)$ and $\alpha_y, \alpha_z \geq \alpha_S(i)$.

For the peer-to-peer relationship, all three clients contribute their resources for media streaming, where x, y, z independently subscribe to i, j, and k stripes from the server, respectively, as shown in Figure 5.25(b). Because y and z are generally out of their respective communication ranges, we let x act as the coordinator. Specifically, y and z periodically send j and k stripes to x, respectively. Together with the media content received from the server, x broadcasts $(i + j + k)$ stripes to y and z. This collaboration arrangement makes each of the three clients effectively subscribe to $(i+j+k)$ stripes, where $(i + j + k) \leq n$. Obviously, the type of x should be higher than the others. Thus, the following condition is necessary: $\alpha_x \geq \alpha_3'(i)$, where $\alpha_3'(i)$ is given by:

$$\alpha_3'(i) = \frac{E_s^{RX}(i) + E_p^{TX}(i + j + k) + E_p^{RX}(j + k)}{E_s^{RX}} \qquad (5.55)$$

On the other hand, y and z have similar type requirements, which depend on the number of subscribed stripes. In particular, the type of y should satisfy: $\alpha_y \geq \alpha_3(j)$, where $\alpha_3(j)$ is given by:

$$\alpha_3(j) = \frac{E_s^{RX}(j) + E_p^{TX}(j) + E_p^{RX}(i + k)}{E_s^{RX}} \qquad (5.56)$$

Similarly, the type of z should satisfy: $\alpha_z \geq \alpha_3(k)$, where $\alpha_3(k)$ is obtained by interchanging the roles of j and k in Equation (5.56). Let us illustrate the peer-to-peer relationship with the following numerical example.

Consider three neighboring clients: x, y, and z whose types are 0.65, 0.45, and 0.45, respectively. If they independently stream the media from the server, x will subscribe to 6 stripes while both y and z will subscribe to 4 stripes. However, they may collaborate and form a peer-to-peer relationship for media streaming, as illustrated in Figure 5.25(b). For y and z, each of them subscribes to 3 stripes from the server and periodically sends them to the coordinator, x. On the other hand, x subscribes to another 4 stripes from the server and periodically broadcasts all media packets to y and z. Effectively, this peer-to-peer relationship allows the three clients to receive 10 stripes, i.e., the complete media content.

5.3.4.4 The General Scenario

Now, let us generalize the analysis to the scenario with r neighboring clients. Because the master-slave relationship does not change with the number of neighboring clients, we focus our attention on the peer-to-peer relationship. Without loss of generality, let us assume that x is the coordinator, which is connected to a set of $(r - 1)$ neighboring helpers, denoted by R_x. In general, these $(r - 1)$ helpers may not be able to communicate with one another. We can denote the number of subscribed stripes by y as s_y, where $y \in R_x$. On the other hand, x subscribes to s_x stripes from the server, where $\sum s \leq n$. We can obtain the threshold value for client x's type, i.e., $\alpha'_r(s_x)$, which is given by:

$$\alpha'_r(s_x) = \frac{E_s^{RX}(s_x) + E_p^{TX}(\sum s) + E_p^{RX}(\sum s - s_x)}{E_s^{RX}} \tag{5.57}$$

Similarly, the threshold value for helper y's type is: $\alpha_r(s_y)$, which is given by:

$$\alpha_r(s_y) = \frac{E_s^{RX}(s_y) + E_p^{TX}(s_y) + E_p^{RX}(\sum s - s_y)}{E_s^{RX}} \tag{5.58}$$

This suggests that a number of neighboring clients satisfying the above thresholds may collaborate to improve the performance of media streaming by utilizing their peer interfaces. This collaboration arrangement allows clients to share the higher energy cost involved in receiving media packets from the server directly. Although the coordinator requires a larger threshold, its energy cost may not be the highest because it may only subscribe to few stripes from the server, i.e., s_x is small or even zero. Because the number of subscribed stripes depends on the client types, heterogeneous clients may form either the master-slave relationship or the peer-to-peer relationship. Motivated by the above analysis, Yeung and Kwok [Yeung and Kwok, 2008] proposed two protocols to guide the establishment of the two relationships.

5.4 Discussion

We have seen that in both wired and wireless systems, the major techniques for providing incentives are: payment (virtual or real), exchange (barter), reciprocity (pairwise), reputation (global), and game theoretic utility. Payment-based systems work by exploiting a user's incentive in increasing or even maximizing its "revenue." However, such an incentive may not be appropriate in some practical situations. For instance, a cellular phone user may not be interested in his or her income (from such sharing) but care more about the quality of service derived from the device. Exchange-based systems fit very well in file sharing applications because users have strong incentives for trading interested files (e.g., music files). For other applications such as forwarding of data (e.g., in wireless ad hoc networks), it is debatable as to whether in practice a user would be interested in exchanging data forwarding capabilities. Reciprocity (in a pairwise manner) is similar in spirit as an exchange-based mechanism. The major difference is that exchange is usually memoryless in that every exchange transaction is treated as a rendezvous event, while a general reciprocity is achieved when devices help each other during different points in time. Thus, such a difference leads to the requirement that users have to keep memory about the prior transactions so that users can "pay back" each other. Due to this history-based feature, reciprocity mechanism suffers from one drawback—a user may be out of the system when he/she needs to pay back. Such a possible "future loss" may deter a user from genuinely contributing to the community for the fear of not getting deserved pay back. Reputation-based systems can be seen as a generalized form of reciprocity. Specifically, while reciprocity is about a particular user pair, a reputation value is a global assessment perceived by all users in the sense that the reputation value is computed by using observations made by many different users. By nature, similar to reciprocity mechanism, reputation systems require substantial memory, centralized or distributed, for recording the reputation values. Thus, it seems that such a system is more suitable for situations where there is a persistent entity, which is logically external to the P2P system, for keeping track of the reputation values. For example, such an entity may be a centralized auctioneer in an electronic auction community, or the base station (access point) in a wireless network. Game theoretic incentive mechanisms are convincing in the sense that the utility function used can usually cover a multitude of important metrics. But the problem is that it is sometimes difficult to achieve an efficient distributed implementation of the resultant protocols.

Perhaps except in an exchange-based system which involves "stateless" interactions, all other incentive mechanisms could suffer from tampering or fabrication of the "incentive parameter" used: the money (virtual or real) in a payment-based system, the reciprocity metric, and the reputation value. Thus, additional mechanisms, usually based on cryptographic techniques, are needed to guard against such potential malicious attacks to the incentive mechanisms. In particular, whitewashing is widely considered as a very low cost technique for a selfish or malicious user to work around the incentive scheme.

Obviously, there is plenty of room for future research about incentive mechanisms in P2P sharing environments. Most notably, revenue maximizing [Yeung and Kwok, 2006b] in a hybrid P2P system (e.g., the so-called converged wireless architecture where an infrastructure-based cellular network is tightly coupled with P2P WLANs) is of a high practical interest because there are more and more cellular subscribers trying to share their resources without the intervention of the cellular service provider. In a data sharing environment, server peer selection [Leung and Kwok, 2005a] is another important direction because we believe that users care more about the quality of service achieved than about the revenue or cost they incur in participation. In economics terms, people, especially wired or wireless game players, are quite inelastic about the costs. Nevertheless, energy conservation [Leung and Kwok, 2005c, Leung and Kwok, 2005b] is still of a prime concern in any wireless P2P sharing network because energy depletion cannot be compensated in any way by increased revenue generated in a payment-based sharing system. Thus, perhaps in a game theoretic setting, we should incorporate energy expenditure in the utility function. Finally, topology control [Leung and Kwok, 2005d] in a wired (overlay) network or wireless (ad hoc) network is also important in the sense that sharing is usually interest based, meaning that users naturally form clusters with similar interests, and as such, related users would be more cooperative in following the incentive protocols. Consequently, building an interest-based sharing topology could be helpful in enhancing the effectiveness of sharing.

5.5 Case Study: PPLive

As in many contemporary practical P2P applications, apparently PPLive does not incorporate any systematic incentive mechanism to promote peer contributions that can possibly lead to an optimized overall performance. Specifically, based on performance studies reported recently [Piatek et al., 2010, Horvath et al., 2008, Vu et al., 2010], it is observed that even a tit-for-tat-like mechanism as that used in BitTorrent clients is not implemented in PPLive. The only premise that such a lack-of-incentive approach can be relied upon is the proprietary nature of the PPLive client programs. Indeed, it is not an exaggeration to say that PPLive is a centralized software system from an implementation point of view because users currently have no control over the client programs' behaviors. Nevertheless, such a situation is bound to change in the near future. More importantly, it is highly probable that system performance can be much enhanced if proper incentive mechanism is in place, as evident by the fact that peers switch channels very often and participation in multiple overlays is far from coordinated.

5.6 Summary

In this chapter, we have presented a detailed survey of incentive techniques for promoting sharing (discrete or continuous data) in a peer-to-peer system. We have considered well-known mechanisms that are proposed and have been deployed on the Internet environment. The techniques used can be classified into: payment based, exchange based, reciprocity, reputation, and game theoretic. These techniques are also applicable in general to a wireless environment. However, a wireless P2P system, while still in its infancy, has a unique challenge—the connectivity among devices is by itself a crucial "sharing" problem (sharing of energy and bandwidth). Much more work needs to be done in related problems such as revenue maximizing (or pricing) in a hybrid wireless P2P system, intelligent peer selection in a data sharing network, energy aware incentive mechanisms, and interest-based topology control. Furthermore, in systems that require payments or some form of critical information items (e.g., reputation values), the security aspect needs to be properly addressed.

Finally, we can see that the amounts of work done in the areas of media streaming and wireless P2P systems are much smaller than that in the area of file sharing. This indicates that these two areas are wide open and are a fertile ground of further research.

5.7 Review Questions

1. What are the different categories of incentive schemes?

2. What are the drawbacks of a payment-based system?

3. What are the main features of BitTorrent's incentive scheme?

4. What are the special incentive requirements for a media streaming system?

5. What are the difficulties involved in enforcing cooperation in a wireless P2P system?

6. What are the adverse effects from the lack of an effective incentive scheme?

Chapter 6

Trust

6.1 Introduction

By nature of a P2P system, trust is a fundamental issue because each peer interacts with another peer which does not have any well-known authority (as in a Web server) and worse still, may "disappear" afterward. Indeed, even for a fully cooperative peer (possibly due to the existence of an incentive system), it is necessary for the peer to determine whether its counterpart in a P2P transaction (e.g., a file downloading operation) is trustworthy or not. For instance, in simple terms, an untrustworthy peer might be one which deliberately injects wrong file data into the network. Consequently, a "good" peer might download some malware from an untrustworthy peer.

There are at least three important issues [Li and Singhal, 2007, Mondal and Kitsuregawa, 2006, Suryanarayana et al., 2005] in the design of a trust management system for P2P networks.

1. We need to quantify "trust" so that each peer can compute the trust values of other peers.

2. We need to specify where the trust values should be stored and maintained. Should we require every peer to store the trust values of every other peer? What about consistency of trust values?

3. We need to come up with a communication protocol for exchanging trust values among peers. More importantly, a proper *aggregation* mechanism must be designed so that each peer can improve the accuracy of its local trust values of other peers by incorporating the trust values sent from remote peers.

In this section, we survey several recently proposed trust management schemes that provide practical answers to the above questions.

6.1.1 Trust Modeling

Azzedin and Maheswaran [Azzedin and Maheswaran, 2003, Azzedin and Maheswaran, 2004] suggested a practical trust computation and management system that is suitable for a P2P computing environment. Specifically, they defined a practical trust model suitable for a P2P computing environment.

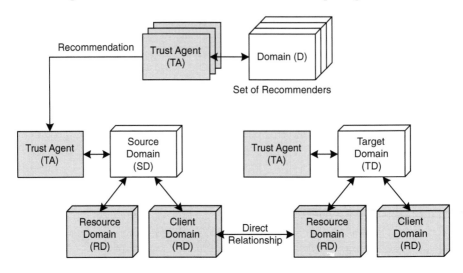

FIGURE 6.1: Trust model for a generic P2P computing system [Azzedin and Maheswaran, 2003].

Figure 6.1 depicts the trust model for a P2P system. The model is partitioned into domains (denoted as D's). There are two virtual domains associated with each D: (1) a resource domain (RD) to signify the resources within the D; (2) a client domain (CD) to signify the clients within the D. Trust agents (TAs) are designated in each D. Each TA has the following functions:

- update the D's trust tables;

- allow entities to join D's and inherit their trust attributes; and

- apply a decay function to reflect the decay of trust between D's.

Each D maintains two data structures: DTT and RTT, which are updated by the TA. The DTT is updated using the trust values observed based on the direct transactions with other D's. The RTT is updated by monitoring the accuracy of recommendations given about target D's.

To illustrate the trust modeling, suppose that an RD in a source domain (SD) is about to establish a trust relationship with a CD associated with a target domain (TD). The SD gathers information to build its direct relations by obtaining its direct relationship TL to TD from its DTT which is internal to the SD. The SD can obtain the reputation value of TD by asking its R. Figure 6.2 shows a recommendation network that enables the establishment of a trust relationship. Each member $z \in R$ provides recommendations based on its DTT. If TD is unknown to recommender z, then z will ask its R. To

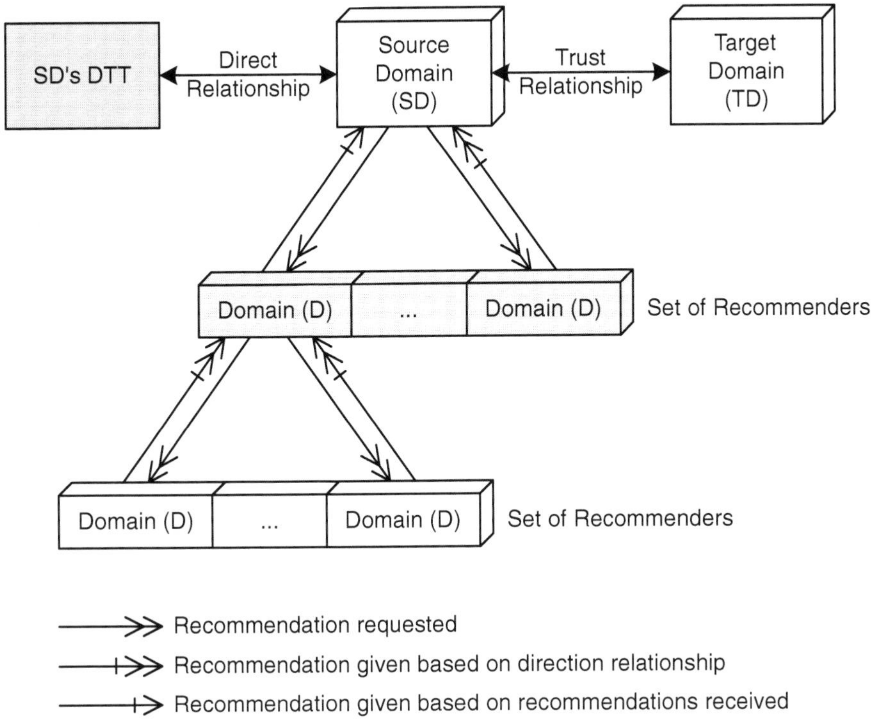

FIGURE 6.2: The trust recommendation mechanism [Azzedin and Maheswaran, 2003].

set up the trust relationship with TD, SD uses two sources of information: (a) the direct trust relationship with TD obtained from SD's DTT, and (b) the reputation trust relationship of TD obtained from SD's R. These two sources of information need to be evaluated and updated if necessary. To update the DTT, a running average of DTT can be kept.

Simulation results indicated that the proposed trust model was not adaptive to inconsistent DTT. Furthermore, if the trust model relies entirely on the direct trust relationship, then it takes a long time to identify the bad domains. However, if the trust model uses both direct relationships and reputation values, identification of bad domains can be more efficient.

6.2   EigenTrust

Kamvar *et al.* [Kamvar et al., 2003] proposed the widely cited EigenTrust system. In EigenTrust, each peer i keeps track of the number of satisfactory transactions it has had with peer j, denoted by $\mathrm{sat}(i,j)$ and the number of unsatisfactory transactions it has had with peer j, denoted by $\mathrm{unsat}(i,j)$. With these parameters, s_{ij} is defined as:

$$s_{ij} = \mathrm{sat}(i,j) - \mathrm{unsat}(i,j) \tag{6.1}$$

Kamvar *et al.* pioneered to point out an insight that the challenge for reputation systems is how to aggregate the local trust values s_{ij} without a centralized storage and management facility. Accordingly, EigenTrust is based on the notion of transitive trust: A peer i will have a high opinion of those peers who have supplied authentic files to it previously. More importantly, peer i is likely to trust the opinions of such peers. Such transitivity of trust is based on the premises that peers who are honest about the files they provide are also likely to be honest in reporting their local trust values.

In quantitative terms, the global reputation of each peer i is given by the local trust values assigned to peer i by other peers, weighted by the global reputations of the assigning peers. Specifically, in EigenTrust, the normalized local trust value, c_{ij}, is defined as:

$$c_{ij} = \frac{\max(s_{ij}, 0)}{\sum_j \max(s_{ij}, 0)} \tag{6.2}$$

However, there are several potential drawbacks in the local trust computation. Firstly, the normalized trust values do not distinguish between a peer with whom peer i did not interact and a peer with whom peer i has had poor experience. Secondly, these c_{ij} values are relative, and there is no absolute interpretation. Specifically, if $c_{ij} = c_{ik}$, we know that peer j has the same reputation as peer k in the opinion of peer i. However, there is no way to tell if both of them are very reputable, or if both of them are just mediocre.

Thus, aggregation of local trust values is necessary. To achieve this, peer opinions are also factored in the weighted evaluation:

$$t_{ik} = \sum_j c_{ij} c_{jk} \tag{6.3}$$

where t_{ik} represents the trust that peer i places in peer k based on asking the trusted peers. Let us denote C as the matrix $[c_{ij}]$ and $\vec{t_i}$ as the vector of values t_{ik}. Thus, the aggregation is given by:

$$\vec{t_i} = C^T \vec{c_i} \tag{6.4}$$

If a peer i carries out this aggregation process for n times, then we have

$\vec{t_i} = (C^T)^n \vec{c_i}$ and it can be shown that [Kamvar et al., 2003] $\vec{t_i}$ will converge to the same vector for all i. Specifically, $\vec{t_i}$ will converge to the left principal eigenvector of C.

Kamvar *et al.* suggested a probabilistic interpretation of EigenTrust: if an agent searches for reputable peers, it can crawl to peer j with probability c_{ij}. After crawling for a while in this manner, the agent is more likely to be at reputable peers than unreputable peers.

In a practical P2P system, there are always a set of enthusiastic peers (i.e., the peers that initiate the system) that warrant unconditional *a priori* trust. To incorporate this feature in the EigenTrust model, Kamvar *et al.* modeled such a set P of pre-trusted peers by a vector $\vec{p_i}$, where each element $p_{ik} = 1/|P|$ if $k \in P$ and $p_{ik} = 0$ otherwise. Thus, each peer i can use this vector to bootstrap the initial trust aggregation: $\vec{t_i}^{(0)} = C^T \vec{p_i}$ and $\vec{t_i}^{(k+1)} = C^T \vec{t_i}^{(k)}$ for all $k \geq 0$.

To combat collusion where a set of malicious peers give each other high reputation values, Kamvar *et al.* suggested a weighted approach:

$$\vec{t}^{(k+1)} = (1-a)C^T \vec{t}^{(k)} + a\vec{p} \qquad (6.5)$$

where a is some constant less than 1. Kamvar *et al.* offered a probabilistic justification for this approach: when a peer is crawling the network, it is less likely to get stuck crawling a malicious member of a collusion because there is a non-zero probability that it ends up at a pre-trusted peer. Another merit of this weighted approach is that the matrix C becomes irreducible and aperiodic, thereby forcing the computation to converge.

Simulation results indicated that using EigenTrust to guide files downloading in a Gnutella-like network can reduce the number of bogus files downloaded under a wide variety of threat scenarios.

However, as shown by Abrams *et al.* [Abrams et al., 2005], the EigenTrust algorithm can also potentially facilitate a selfish peer to lie about its recommendation to gain a higher trust value. Specifically, to maximize its trust, a peer must always recommend a peer that recommended it. Consider the download graph of Figure 6.3(a), and assume a uniform distribution for pre-trusted peers over all n peers: if the middle peer reports a download from the right peer, it will have trust $\frac{(2-\epsilon)\epsilon}{n}$. If, on the other hand, it reports a download from the left peer, it will only have a trust value of $1/n$.

Abrams *et al.* [Abrams et al., 2005] suggested a modified EigenTrust algorithm which works by first creating a query topology and then making each peer's trust independent of his/her reporting of downloads. Specifically, the EigenTrust algorithm is modified as follows:

Initialization At the initialization of the modified algorithm, peers are partitioned into groups evenly (depicted by different colors), where $C = \{c_1, c_2, \ldots, c_m\}$ is the set of partitions. Each color has either $\lfloor \frac{n}{m} \rfloor$ or $\lceil \frac{n}{m} \rceil$ peers. The colors are arranged into a directed cycle chosen uniformly at random. Then, $\forall c \in C$, let $\text{pred}(c)$ be the color which is the

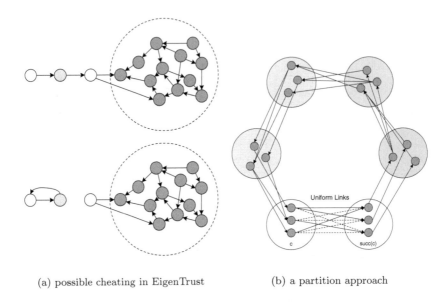

(a) possible cheating in EigenTrust   (b) a partition approach

FIGURE 6.3: Drawback of EigenTrust and a modified situation [Abrams et al., 2005].

predecessor of c in the cycle and succ(c) the successor of c. The distribution p over pre-trusted peers is restricted in such a way to assign an equal amount of pre-trusted weight to each color.

Run Transactions Each peer i in every color c is allowed to query and download only from peers in succ(c). Thus, for every query q, the set of servers contains only peers in succ(c).

Compute Trust Values In order to compute the trust score for nodes of a given color c, the stationary distribution of a modified Markov chain is determined. The outgoing links from color c are set to be uniform over succ(c), and then the trust values of the nodes in c in this modified Markov chain are computed as shown in Figure 6.3(b).

Simulation results indicated that the improved EigenTrust algorithm performed almost as well as the original EigenTrust but demonstrated better load balancing properties.

6.3 PeerTrust

Xiong and Liu [Xiong and Liu, 2004] proposed the PeerTrust mechanism. In PeerTrust, similar to prior approaches such as EigenTrust, a peer's trustworthiness is defined by an evaluation of the peer it receives in providing service to other peers in the past. Xiong and Liu identify five important factors for such evaluation:

1. the feedback a peer obtains from other peers;

2. the feedback scope, such as the total number of transactions that a peer has with other peers;

3. the credibility factor for the feedback source;

4. the transaction context factor for discriminating mission-critical transactions from less or noncritical ones; and

5. the community context factor for addressing community-related characteristics and vulnerabilities.

Xiong and Liu's approach is based on the notation as shown in Table 6.1.

TABLE 6.1: Notation used in Xiong and Liu's PeerTrust system.

Symbol	Definition
$I(u,v)$	total number of transactions performed by peer u with v
$I(u)$	total number of transactions performed by peer u with all other peers
$p(u,i)$	other participating peer in peer u's ith transaction
$S(u,i)$	normalized amount of satisfaction peer u receives from $p(u,i)$ in its ith transaction
$Cr(v)$	credibility of the feedback submitted by v
$TF(u,i)$	adaptive transaction context factor for peer u's ith transaction
$CF(u)$	adaptive community context factor for peer u

Note: From Xiong and Liu, 2004.

The trust value of peer u denoted by $T(u)$ is then defined as:

$$T(u) = \alpha \sum_{i=1}^{I(u)} S(u,i) Cr(p(u,i)) TF(u,i) + \beta CF(u) \qquad (6.6)$$

where α and β denote the normalized weight factors for the collective evaluation and the community context factor. The first part of the trust computation equation is a weighted average of amount of satisfaction a peer receives for each transaction. The second part scales the first part by an increase or decrease of the trust value based on community-specific characteristics and situations.

For the first part, there are two variations in defining the credibility measure. The first one is to use a function of the trust value of a peer as its credibility factor. Thus, feedback from trustworthy peers is considered more credible and, consequently, weighted more than that from untrustworthy peers. This definition of credibility measure is based on two assumptions: (1) untrustworthy peers are more likely to submit false or misleading feedback in order to cover up their own malicious behavior; (2) trustworthy peers are more likely to be honest on the feedback they provide. Accordingly, considering only the first component, the trust metric is now given by:

$$T_{TVM}(u) = \sum_{i=1}^{I(u)} S(u,i) \frac{T(p(u,i))}{\sum_{j=1}^{I(u)} T(p(u,j))} \tag{6.7}$$

The second possible credibility measure is for a peer w to use a personalized similarity measure to rate the credibility of another peer v through w's prior interactions experience. Specifically, peer w uses a personalized similarity between itself and another peer v to weight the feedback by v on any other peers. Let $IS(v)$ denote the set of peers that have interacted with peer v. Thus, the common set of peers that have interacted with both peer v and w, denoted by $IJS(v,w)$, is given by $IS(v) \cap IS(w)$.

To measure the feedback credibility of peer v, peer w computes the feedback similarity between w and v over the common set $IJS(v,w)$ of peers that they have interacted with in the past. Here, the feedback by v and the feedback by w over $IJS(v,w)$ are modeled as two vectors. As a result, the credibility can be defined as the similarity between the two feedback vectors. The root-mean-square or standard deviation (dissimilarity) of the two feedback vectors can then be used to compute the feedback similarity. Accordingly, the trust metric (considering only the first component) is given by:

$$T_{PSM}(u,w) = \sum_{i=1}^{I(u)} S(u,i) \frac{Sim(p(u,i),w)}{\sum_{j=1}^{I(u)} Sim(p(u,j),w)} \tag{6.8}$$

where:

$$Sim(v,w) = 1 - \sqrt{\frac{\sum_{x \in IJS(v,w)} \left(\frac{\sum_{i=1}^{I(x,v)} S(x,i)}{I(x,v)} - \frac{\sum_{i=1}^{I(x,w)} S(x,i)}{I(x,w)} \right)^2}{|IJS(v,w)|}} \tag{6.9}$$

Figure 6.4(a) gives a sketch of the system architecture of PeerTrust. First of all, we can see that there is no central database, implying that trust data that are needed to compute the trust measure for peers are stored across the network in a fully distributed manner. The trust manager performs two main functions. Firstly, it submits feedback to the network through the data

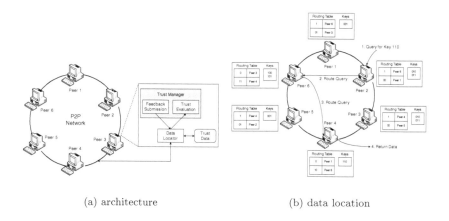

(a) architecture (b) data location

FIGURE 6.4: System architecture and data location mechanism in PeerTrust [Xiong and Liu, 2004].

locator, which routes the data to appropriate peers for storage. Secondly, it is responsible for evaluating the trustworthiness of a particular peer. This task is performed in two steps. It first collects trust data about the target peer from the network through the data locator and then computes the trust value.

Trust data location is based on P-Grid [Aberer, 2001]. As shown in Figure 6.4(b), the trust data about a peer u, i.e., feedback u receives for each transaction, are stored at designated peers that are located by hashing a unique ID of peer u to a data key. Each piece of feedback includes the following information: ID of peer u as the data key, timestamp, or counter of the transaction, feedback for that transaction, ID of the peer who provides feedback, and other applicable transaction contexts. Each peer is responsible for multiple keys and maintains a routing table for other keys. When a peer receives a search or update request with a data key that it is not responsible for, it forwards the request according to its routing table. Consequently, the storage cost at each peer is proportional to the degree of replication and the amount of history information data that it needs to store.

Simulation results indicated that PeerTrust is effective in combating collusion attacks and has a reasonably low error in trust computations.

6.4 Trust-χ

Motivated by the observation that many previous work in trust modeling mainly focused on only one aspect of trust negotiation such as policy and credential negotiation, or the selection of the negotiation policy, Bertino *et al.* [Bertino et al., 2004] proposed a comprehensive solution called Trust-χ, an XML-based system addressing all the phases of a negotiation and providing novel features with respect to existing approaches.

The first component of Trust-χ is an XML-based language, named χ-TNL, for composing certificates and policies. Trust-χ certificates are either credentials or declarations. Here, a credential states personal characteristics of its owner, certified by a Credential Authority (CA), whereas declarations collect personal information about its owner that do not need to be certified (such as, for instance, specific preferences) but may help in better customizing the offered service. A novel aspect of χ-TNL is the support for trust tickets, which are issued upon the successful completion of a negotiation and can be used to speed up subsequent negotiations for the same resource. Additionally, χ-TNL allows the specification of a wide range of policies and provides a mechanism for policy protection, based on the notion of policy preconditions.

A Trust-χ negotiation consists of a set of phrases to be sequentially executed. A salient feature of Trust-χ is that it provides a variety of strategies for trust negotiations, which allow a peer to better trade off between efficiency and protection requirements. The motivation behind this design is that, since trust negotiations can be executed for several types of resources and by a variety of entities having various security requirements and needs, a single approach to perform negotiation processes may not be adequate in all the circumstances. As a result, Trust-χ is very flexible and can support negotiations in a variety of scenarios, involving entities like business, military and scientific partners, or companies and their cooperating partners or customers.

As shown in Figure 6.5, according to the design rationales underlying Trust-χ, each entity in the P2P system can be the controller of one or more resources, a third-party credential issuer, or a requester. Typically, a negotiation involves two entities: the entity providing negotiated resources, referred to as the controller, and the entity wishing to access the resources, referred to as requester. Note that the controller does not necessarily coincide with the owner of the resource, it may be the manager of the resource entitled by the real owner.

Each entity, characterized by a Trust-χ profile of certificates, can act as a requester in one negotiation and as a controller in another. During a negotiation, mutual trust might be established between the controller and the requester. Specifically, the requester has to show its certificates to obtain the resource, and the controller, whose honesty is not always assured, submits certificates to its counterpart in order to establish trust before receiving sensitive information. Release of information is regulated by disclosure policies, which are exchanged to inform the other party of the trust requirements that need to be satisfied to advance the state of the negotiation. Trust-χ participants are both considered equally important. Thus, each party has an associated system managing negotiation and always has a complete view of the state of the negotiation process.

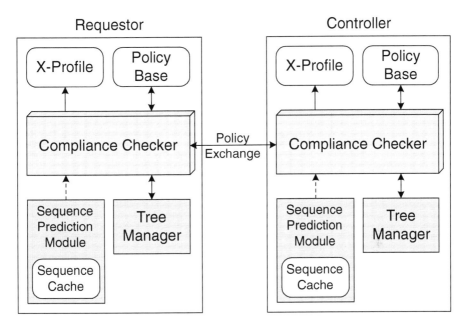

FIGURE 6.5: Trust-χ architecture [Bertino et al., 2004].

The system is also composed of a Policy Base, storing disclosure policies, the χ-Profile associated with the party, a Tree Manager, storing the state of the negotiation, and a Compliance Checker, to test policy satisfaction and determine request replies. The Compliance Checker includes a credential verification module, which performs a validity check of the received credentials in order to verify the document signature, check for credential revocation, and discovery credential chain, if necessary. Finally, Trust-χ system has a complementary module named Sequence prediction module, for caching and managing previously used trust sequences.

6.5 FuzzyTrust

Based on a detailed analysis on the characteristics of eBay's transaction data, Song *et al.* [Song et al., 2005] developed a FuzzyTrust prototype system for evaluating peer reputation in P2P transactions. The FuzzyTrust is built with a fuzzy logic inference technique (elaborated below). A salient and novel feature of FuzzyTrust is that the system is capable of handling imprecise or uncertain information collected from the peers.

To explain basic fuzzy concepts, let us consider the seller's local score inference example. In fuzzy theory, the membership function $\mu(x)$ for a fuzzy variable x specifies the degree of an element belonging to a fuzzy set. It maps x into the range $[0, 1]$, where 1 is full membership and 0 is no membership. Figure 6.6(a) shows a high membership function for modeling the local score (Γ), and Figure 6.6(b) shows the five levels of membership function. Figure 6.6(c) illustrates the inference process. Consider two fuzzy variables: one is the product quality (Q) and another is the delivery time (T), with initial values $Q = 0.84$ and $T = 0.26$. To illustrate, we apply the following two simple fuzzy inference rules in Figure 6.6:

1. If Q is very good AND T is moderate, then *Gamma* is high.

2. If Q is ordinary AND T is fast, then *Gamma* is medium.

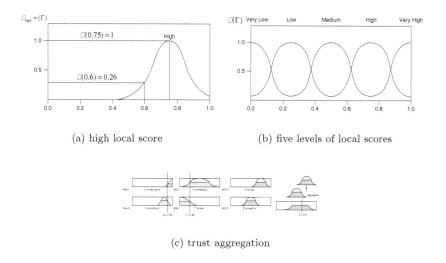

(a) high local score   (b) five levels of local scores

(c) trust aggregation

FIGURE 6.6: Illustration of fuzzy inference [Song et al., 2005].

All rules are processed in parallel. The resulting membership is determined by assessing all terms in the premise. The fuzzy operator AND is then applied to determine the support degree of the rules. The AGGREGATE operator superimposes two resulting membership curves. The final local score $\Gamma = 0.6$ is generated by defuzzifying from the aggregation result. This is done by taking the centroid of the superimposed membership curve in Figure 6.6(c). In a real-life P2P reputation system, this fuzzy logic inference process could demand tens to hundreds of rules.

In FuzzyTrust, three system design criteria are developed below to match with the eBay characteristics.

1. The network bandwidth consumption required to exchange local trust scores for hot spots can be extremely high. Thus, a reputation system

for e-transactions should consider the unbalanced transactions among users.

2. Second, to address the lesser impact from small users, a reputation system should not apply the same evaluation cycle for all peers. The super users should be updated more often than the small users.

3. Third, with skewed transaction amount, it makes sense to evaluate the large transactions more often than the small ones.

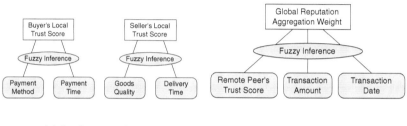

(a) local trust inference   (b) global trust aggregation

FIGURE 6.7: Local and global trust management mechanisms in FuzzyTrust [Song et al., 2005].

As shown in Figure 6.7, the FuzzyTrust system works by performing two major inference steps: local-score calculation and global reputation aggregation.

Local-Score Calculation. In FuzzyTrust, peers perform fuzzy inference on local parameters to generate the local scores. Figure 6.7(a) illustrates the local-score calculation for eBay transactions. The fuzzy logic sidebar shows a detailed fuzzy inference procedure. The fuzzy inference mechanism can capture some uncertainties and is self-adjusting. It can adaptively track the variation of local parameters such as payment method and time, goods quality, and delivery time, etc.

Global Reputation Aggregation. The FuzzyTrust system must then aggregate local trust scores collected from all peers to produce a global reputation for each peer. The system uses fuzzy inference to obtain the global reputation aggregation weights, as illustrated in Figure 6.7(b). The aggregation weights are determined by three variables: the peer's reputation, transaction date, and transaction amount.

Listed below are several frequently used fuzzy inference rules in the prototype FuzzyTrust system construction. In a full-scale P2P reputation system, the number of fuzzy inference rules could be extended to several hundreds.

1. If the transaction amount is very high and the transaction time is new, then the aggregation weight is very large.

2. If the transaction amount is very low or the transaction time is very old, then the aggregation weight is small.

3. If a peer's reputation is good and the transaction amount is high, then the aggregation weight is very large.

4. If a peer's reputation is good and the transaction amount is low, then the aggregation weight is medium.

5. If a peer's reputation is bad, then the aggregation weight is very small.

FuzzyTrust then computes the global reputation using the following definition:

$$R_i = \sum_{j \in S} \left(\frac{w_j}{\sum_{j \in S} w_j} t_{ji} \right) \tag{6.10}$$

where R_i is the global reputation of peer i, S is the set of peers with whom peer i has conducted transactions, t_{ji} is the local trust score of peer i rated by peer j, w_j is the aggregation weight of t_{ji}. The global aggregation process runs multiple iterations until each R_i converges to a stable global reputation rating for peer i.

Song *et al.* implemented the prototype FuzzyTrust system on a DHT-based P2P overlay network, with an architecture similar to that of Chord. This DHT ring provides fast trust aggregation and secure message transmission. The Chord system is highly scalable, robust to failure, and self-organizing in that it handles peer join and leave from the system. Figure 6.8 shows the DHT-based FuzzyTrust system architecture.

Each peer maintains two tables: a transaction record table to maintain transaction records with remote peers, and a local score table to maintain remote peers' evaluated trust scores. Based on the transaction records, FuzzyTrust infers the global aggregation weights through the fuzzy inference system. When performing global reputation aggregation, each peer queries the trust scores from remote peers. To tackle the hot-spot issue, the system partially queries qualified peers that meet an aggregation threshold. Figure 6.9 shows an example of global reputation aggregation based on the DHT configuration in Figure 6.8.

In Song *et al.*'s simulation study based on eBay trace data, FuzzyTrust consistently outpeformed EigenTrust in terms of convergence time, error in detecting malicious users, and message overheads.

Griffiths *et al.* [Griffiths et al., 2006] extended Song *et al.*'s FuzzyTrust framework by incorporating a new notion called undistrust. Griffiths *et al.* observed that most previous work on trust has concentrated on the positive aspect of trust but largely ignored the notion of distrust.

Griffiths *et al.* insightfully observed that distrust is not simply the negation

Peer 2 Transaction Record Table

Remote Peer ID	Remote Peer's Trust Score	Transaction Amount	Transaction Date	Global Aggregation Weight
4	0.5	$15	02/11/2005	0.5
9	0.7	$10	02/15/2005	0.6
20	0.9	$99	02/13/2005	0.8
28	0.8	$399	02/14/2005	0.9

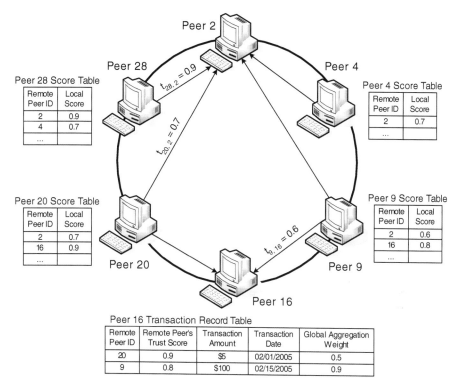

Peer 28 Score Table

Remote Peer ID	Local Score
2	0.9
4	0.7
...	

Peer 4 Score Table

Remote Peer ID	Local Score
2	0.7
...	

Peer 20 Score Table

Remote Peer ID	Local Score
2	0.7
16	0.9
...	

Peer 9 Score Table

Remote Peer ID	Local Score
2	0.6
16	0.8
...	

Peer 16 Transaction Record Table

Remote Peer ID	Remote Peer's Trust Score	Transaction Amount	Transaction Date	Global Aggregation Weight
20	0.9	$5	02/01/2005	0.5
9	0.8	$100	02/15/2005	0.9

FIGURE 6.8: Illustrative example of FuzzyTrust [Song et al., 2005].

of trust but rather it is a belief that a peer will act against the best interests of another. Alternatively, untrust corresponds to the space between distrust and trust, in which an agent is positively trusted, but not to the extent that it warrants full cooperation. This view of trust was originally advocated by Marsh and Dibben [Marsh and Dibben, 2005]. Inspired by this definition, Griffiths *et al.* suggested a novel concept called undistrust.

According to Griffiths *et al.*, a similar region of undistrust is needed, namely negative trust, but insufficient to make definite conclusions in the trust reasoning process.

Figure 6.10(b) illustrates their definition of the notions of trust, distrust, untrust, and undistrust.

Preliminary results presented in [Griffiths et al., 2006] indicated that the

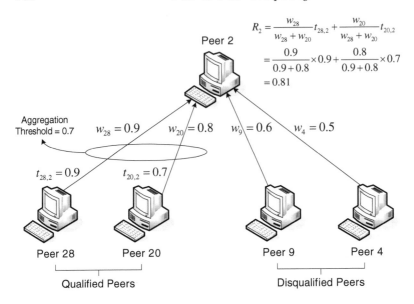

FIGURE 6.9: Example of global trust aggregation [Song et al., 2005].

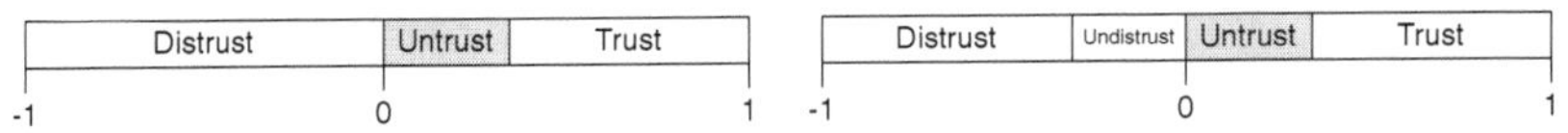

| Distrust | Untrust | Trust | | Distrust | Undistrust | Untrust | Trust |

(a) Marsh and Dibben's definition   (b) Griffiths *et al.*'s new notion

FIGURE 6.10: Notions of trust [Griffiths et al., 2006].

proposed new notion of undistrust can generate better performance in terms of errors in identifying malicious peers.

6.6   Game Theoretic Analysis on Trust Management

Recently, Tuan [Tuan, 2006] presented an interesting and insightful game theoretic analysis of a general trust management system. Tuan started by modeling the trust reporting process as a *mixed strategy* game [Osborne, 2004], and found that if a reputation system is not incentive-compatible, the more the number of peers in the system, the less likely that anyone will report about a malicious peer.

Specifically, Tuan made use of a reasonable assumption that the cost of reporting a malicious peer might be very high. Specifically, the cost of reporting might be time and/or the risk of possible retaliation. To model the situation, Tuan assumed that there exists a certain level of incentive for each peer to report about a malicious peer. However, peers would prefer someone else to do that (i.e., she is better off if other peers do the reporting). This is a typical "reporting a crime" situation [Osborne, 2004] where many people witnessed a crime but any one of them is unwilling to report the crime [Rosenthal, 1964].

In Tuan's model, assume that each peer is satisfied if a malicious peer is reported and attaches a value s to this. Reporting is costly and the cost is assumed to be c, where $s > c > 0$. Thus, there are three possible cases for each peer: (1) the malicious peer is not reported and the payoff is 0; (2) the malicious peer is reported by the peer and thus the payoff is $s - c$; and (3) the malicious peer is reported by some other peer and thus the payoff is s.

Consider a mixed strategy situation where each peer probabilistically chooses to report or not. Denote the probability that each peer would report as p. Given a peer, the probability that no one out of $(n - 1)$ remaining peers reports is thus $(1 - p)^n$. Consequently, the probability that at least one peer (out of the remaining $(n - 1)$ peers) reports is $1 - (1 - p)^{n-1}$. By definition, in equilibrium state [Osborne, 2004] the expected payoff of reporting for each peer is equal to the expected payoff of not reporting. Thus, we have:

$$s - c = 0 + s(1 - (1 - p)^{n-1}) \tag{6.11}$$

Solving this equation gives:

$$p = 1 - \left(\frac{c}{s}\right)^{\frac{1}{n-1}} \tag{6.12}$$

Now, we can see that the probability that each peer reports about the malicious peer decreases as the number of reporting peers increases.

Furthermore, Tuan also addressed the issue of voting for exclusion of a (maliciously believed) peer and provided an analysis of the problem. By modeling the decision process as a Bayesian game [Osborne, 2004], Tuan found that the possible application of exclusion in P2P system might be dangerous. In Tuan's Bayesian game model, the voting super peers have some *a priori* belief about the type of the peer in question. Now, depending on the availability and accuracy of new information about such a suspicious peer, this belief can be changed. Tuan's analysis showed that, under certain assumptions, the more the number of voting peers, the more likely that an innocent peer is excluded from the network.

6.7   SuperTrust

Dimitriou *et al.* [Dimitriou et al., 2007] recently proposed an interesting scheme called SuperTrust. A novel feature of SuperTrust is that the trust reports are encrypted and are never opened during the submission or aggregation processes, thus guaranteeing privacy, anonymity, fairness, persistence, and eligibility of transactions.

SuperTrust is a decentralized framework that ensures the security of trust handling in K-redundant super peer networks. Thus, in some sense, SuperTrust is orthogonal to existing trust management systems for ordinary peers. However, SuperTrust relies on a hybrid network architecture in that it assumes the existence of some certificate authority (CA) that can generate or certify special purpose keys and whose public key can be trusted as authentic.

Associated with each peer v in SuperTrust is a chosen set of n super peers (i.e., aggregators) that are responsible for "collecting" the votes/reports of other peers that have interacted with v. The aggregators for each peer are chosen by the CA amongst the K super peers responsible for the various clusters. Furthermore, in each cluster, the CA delegates a storage node chosen amongst the K super peers to act as a storage facility for the reputations of the peers/resources located in the corresponding cluster (alternatively, as suggested by Dimitriou *et al.*, this role can be assumed by the aggregators, thus eliminating single point of failure in the system).

Such a semi-centralized, semi-distributed approach guarantees that each aggregator peer is within a fixed number of hops from each peer, thereby potentially improving the overall performance of the system. The various actions of a peer v in SuperTrust are outlined below (see Figure 6.11) [Dimitriou et al., 2007]:

Step 1: Send a file request. Peer v isssues a request for resource r. Upon reception of v's request, one of the super peers responsible for v's cluster broadcasts this request to their neighbors.

Step 2: Receive a list of relevant peers. Upon reception of v's request, each super peer checks whether the resource requested is within its cluster. Peer u issues a reply confirming his/her possession of the requested resource. In addition, each of n aggregators of u partially decrypts the encrypted trust value of u using a (t, n) Paillier-based threshold cryptosystem [Paillier, 1999], and responds to v with their decrypted shares allowing v to compute the final trust value.

Step 3: Select a set of peers. Once peer v receives the replies and the decrypted shares from a sufficient number t of aggregators, it calculates the global trust value of the replying peers and chooses to download the resource from the most reputable peer.

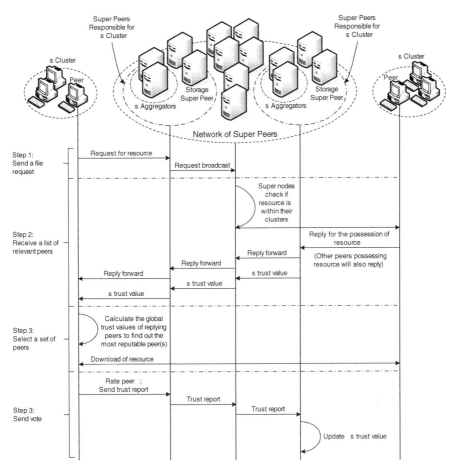

FIGURE 6.11: Example of secure trust aggregation in SuperTrust [Dimitriou et al., 2007].

Step 4: Send vote. Peer v rates the interaction it had with peer u. It first encrypts the report with the public key of the u's aggregators. Doing this allows encapsulating its rating for both peer u and its resource. The peer then submits it to the designated super peer. The latter forwards the encrypted vote to its neighbors. Upon reception of the trust report for v, the aggregators fetch from the storage super peer v's previous encrypted ratings, update it using a homomorphic encryption scheme [Paillier, 1999], and submit the aggregation result back to the storage Super peer in order to guarantee the durability of the ratings in the system. In turn, the storage super peer only stores the encrypted value that was advertised by the majority of the aggregators. Here, we can see that such a scheme protects against up to $n/2$ suspicious aggregators (where n is the total number of aggregators in some cluster) that are trying to cheat the system by submitting erroneous aggregation results. Now, the global trust value of v is updated by v's aggregators without decrypting the intermediate reports, thereby ensuring privacy and integrity of votes.

Simulation results indicated that SuperTrust performed well in terms of message overhead and response time.

6.8 PowerTrust

Zhou and Hwang proposed a DHT-based trust management scheme called PowerTrust [Zhou and Hwang, 2007b]. Figure 6.12(a) shows the system architecture of PowerTrust. First, a trust overlay network (TON) is built on top of all peers in the P2P system. All peers send local trust scores among themselves periodically. The PowerTrust system then aggregates the local scores to compute the global reputation score of each participating peer. All global scores form a reputation vector $V = (v_1, \ldots, v_n)$, where $\sum_i v_i = 1$.

As shown in Figure 6.12(a), the regular random walk module supports the initial reputation aggregation. The lookahead random walk (LRW) module is used to update the reputation score periodically. To this end, the LRW also works with a distributed ranking module to identify the power peers (i.e., peers with prior trust). The system leverages the power peers to update the global reputation scores.

In PowerTrust, feedback scores are generated by Bayesian learning or by an average rating based on peer satisfaction. Each peer normalizes all issued feedback scores. Consider the trust matrix $R = (r_{ij})$ defined over the n-peer TON, where r_{ij} is the normalized local trust score defined by $r_{ij} = \frac{s_{ij}}{\sum_j s_{ij}}$, and s_{ij} is the most recent feedback score that peer i rates peer j. If there is no link from peer i to peer j, s_{ij} is set to 0. Thus, for all $1 \leq i, j \leq n$, we have $0 \leq r_{ij} \leq 1$ and $\forall i$, $\sum_{j=1}^{n} r_{ij} = 1$. In other words, matrix R is a stochastic matrix in which all entries are fractions and each row sums to 1. This requires that the scores issued by the same peer to other peers are normalized.

The reputation vector is then recursively updated as follows:

$$V_{(t+1)} = R^T \times V_{(t)} \tag{6.13}$$

This updating is done until $|V_{(i)} - V_{(i-1)}| \leq \epsilon$.

Using the parameter sweeping P2P Grid applications in their simulation study, Zhou and Hwang found that PowerTrust performed well in terms of trust aggregation accuracy, message overhead, and makespan of the Grid applications.

6.9   GossipTrust

Recently Zhou and Hwang also proposed another novel trust management
scheme, called GossipTrust [Zhou and Hwang, 2007a], designed for unstructured P2P systems such as KaZaA. Figure 6.12(a) shows the architecture of
GossipTrust.

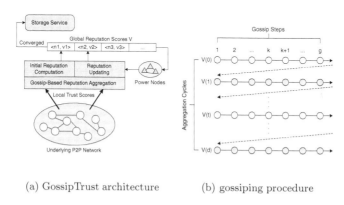

(a) GossipTrust architecture   (b) gossiping procedure

FIGURE 6.12: GossipTrust [Zhou and Hwang, 2007b, Zhou and Hwang,
2007a].

In GossipTrust, each peer keeps a row vector of trust matrix S based
on its outbound local trust scores. In addition, each peer also maintains a
global reputation vector $V(t)$ at aggregation cycle t. Internally, this vector is
represented by a collection of <nodeID, score> pairs. At the first aggregation
cycle, $V(0)$ is initialized with equal global reputation scores, i.e., $v_i(0) = 1/n$,
for $i = 1, 2, \ldots, n$.

To compute the successive reputation vectors, GossipTrust uses a gossip-
based protocol to perform the matrix-vector computation. Gossiping supports
lightweight communications among nodes during the aggregation process. In
GossipTrust, each aggregation cycle consists of several gossip steps as shown
in Figure 6.12(b). In a gossip step, each peer receives reputation vectors from
others, selectively integrates the vectors with its current reputation vector, and
then sends the updated one to a random peer in the network. This gossiping
process continues until the gossiped scores converge in g steps, where g is
determined by a set of gossiping error threshold ϵ. After the convergence
of gossip steps, GossipTrust continues the next aggregation cycle until the
global reputation vectors converge in d cycles, where d is determined by the
aggergation error threshold δ.

Simulation results indicated that GossipTrust generated good performance in terms of trust aggregation accuracy, convergence rate, and message overhead.

6.10   Trust Establishment in Wireless Sensor Networks

As discussed in Chapter 2, wireless sensor networks (WSNs) are considered practical P2P distributed processing platforms for many applications (e.g., battle-field data processing). Trust establishment has long been a challenging problem in WSNs due to high security requirements and strict resource constraints in WSNs. Recent approaches based on randomized key pre-distribution schemes mainly focus on key allocations supported by the pre-deployment estimations of the post-deployment information items. Unfortunately, such information items may be unavailable or may change over time. The performance of the resulting network may be unstable and unable to react to the change of topology due to sensor node dynamics.

In the following, we present a detailed survey of recently proposed techniques for trust establishment in WSNs.

6.10.1   Symmetric Key-Based Approaches

Generally, these schemes are further divided into two categories: *deterministic* and *probabilistic* key pre-distribution schemes.

6.10.1.1   Deterministic Key Pre-Distribution Schemes

A deterministic scheme assigns keys to each node intentionally so that these keys are used for specific purposes. Since each key is selected carefully, it is expected that the number of keys stored for each node will not be large; otherwise, the cost of key assignment will be very high if the network is deployed in a large scale. Thus, this scheme usually focuses on the communication between a fixed number of entities such as node-to-base station, node-to-gateway, and gateway-to-base station. A simple example is the master key approach in which each node communicates with a single common key [Kwok, 2007].

Jolly *et al.* [Jolly et al., 2003] proposed a lightweight key management protocol. They observed that sensor-to-sensor secure communication is not always necessary in some applications. Thus, very few keys (typically two) are enough to establish secure connections between nodes and base station as well as cluster gateways. Since there is no sensor-to-sensor communication, the first key is used to talk with the base station while the second one is used to talk with the cluster gateway. These two keys can be computed efficiently and distributed to sensor nodes before deployment. Their objective is to provide a cost-effective key infrastructure to secure the sensor network. However, their design requires a centralized key server and the re-keying[1] process suggested is inefficient due to the large amount of message exchanges (i.e., a large overhead).

Based on a trusted intermediary node and the underlying routing protocol, Chan and Perrig [Chan and Perrig, 2005] introduced a class of key establishment protocols, called Peer Intermediaries for Key Establishment (PIKE). By assuming the existence of routing information, PIKE uses a third node C located somewhere in the network to act as a trusted intermediary between two nodes A and B. The trusted entity shares a common key with both node A and node B so that the key establishment protocols can be securely routed through node C to perform connection establishment. The objectives of PIKE are to provide a uniform communication pattern for key establishment and reduce the communication and memory overheads when the network size increases. However, the dependence on the underlying routing protocol makes this scheme less attractive and it is hard to adapt to topology changes. It is noted that PIKE is considered as a deterministic key pre-distribution scheme because any two nodes are guaranteed to be able to set up a key.

6.10.1.2 Probabilistic Key Pre-Distribution Schemes

Eschenauer and Gligor [Eschenauer and Gligor, 2002] proposed a random key pre-distribution scheme (referred to as *basic scheme* in this chapter) in 2002. Based on random graph theory [Erdös and Rényi, 1960], the *basic scheme* relies on probabilistic key sharing among nodes and uses a simple shared key discovery protocol and path key establishment for the connection setup process. This scheme assumes that the sensor network forms a random graph and keys are installed in nodes prior to deployment. Each sensor node installs a random set of keys from the key pool. Any two neighbors are connected if they are able to find a common key.

The principle of key pre-distribution is widely adopted in many key management schemes in WSNs. One of the major reasons is that it can provide an acceptable level of security on the resource-constrained sensor nodes. After the pioneering work of the random key pre-distribution scheme proposed by Eschenauer and Gligor [Eschenauer and Gligor, 2002], many enhancements on the *basic scheme* have been proposed. In the following, we briefly discuss several typical trust establishment schemes which enhance the *basic scheme* in different ways.

Chan *et al.* [Chan et al., 2003] proposed a q-composite key pre-distribution scheme in which q common keys ($q > 1$) are required to establish a single secure link between two neighboring nodes. This scheme achieves better security under small scale attack (when the fraction of compromised nodes is less than 0.5%) while increasing the vulnerability when the number of compromised nodes increases. They also proposed a *random pairwise scheme* which can

---

[1]Re-keying refers to the process of replacing an old key with a new one.

be regarded as a randomized version of the pairwise scheme. In the pairwise scheme, each sensor node has $n-1$ keys which are privately shared with another node in the network. To reduce the memory storage, the random pairwise scheme only picks a subset of keys from those $n-1$ keys. Thus, the memory usage of the sensor node is reduced but the network connectivity is also decreased. The random pairwise scheme provides full resilience against node capture attack as even when some nodes are compromised, the remainder of the network remains fully secure. However, due to the limited memory storage of sensors, each node can only hold a limited amount of unique keys. The number of keys stored cannot scale well with the increasing network size. As a result, the maximum network size supported is smaller than that of the *basic scheme*.

Based on the known attack probabilities[2] in different regions, Chan *et al.*'s scheme [Chan et al., 2005b] targets at enhancing the overall network resilience[3]. Their scheme adjusts the number of distinct keys stored in a sensor node depending on the attack probability of the region they are going to be deployed. If a node is going to be deployed in a region with a higher attack probability, fewer keys will be assigned to it and vice versa. In this case, the adversaries have a higher probability to attack a node with fewer keys stored, and therefore the number of keys exposed to the adversaries after each attack is smaller. It is clear that there are two drawbacks in this approach. Firstly, a certain extent of connectivity is sacrificed. Secondly, additional information on a known and fixed attack probability is required to be known before deployment. Nevertheless, the authors showed that there is a substantial improvement in network resilience.

Du *et al.* [Du et al., 2005] proposed a random key pre-distribution method built on the Blom key pre-distribution scheme [Blom, 1984] to improve network resilience. Blom's original scheme uses a single key space to allow any pair of nodes to compute a secret key. Each node is required to store $\lambda+1$ keys and the scheme guarantees that as long as no more than λ nodes are compromised, all the links between non-compromised nodes remain secure. Du *et al.* extended this idea to multiple key spaces instead of a single one. Two nodes share a pairwise key only if they hold a common key space. Their scheme keeps the λ-secure property but relaxes the memory requirement. Consequently, the network formed results in a *connected* graph with a probability instead of a guaranteed *complete* graph by Blom's scheme. Later, Du *et al.* [Du et al., 2004] tried to reduce the memory usage using the deployment knowledge (e.g., location) while achieving the same level of connectivity. However, such a piece of knowledge is not always available, especially in a hostile area and dynamic network environment. Specifically, offline estimation of the node distribution

---

[2]Attack probability refers to the probability that a node within a certain region will be captured.

[3]Network resilience generally refers to the ability of a network to resist being affected by some attacks.

after deployment is generally considered as unrealistic and inaccurate under these situations.

Liu *et al.* [Liu et al., 2005a] proposed a polynomial pool-based pairwise key pre-distribution scheme. Instead of deploying keys, their scheme uses bivariate t-degree polynomials $f(x, y) = \sum_{i,j=0}^{t} a_{ij} x^i y^j$, such that $f(x, y) = f(y, x)$, to compute the pairwise communication keys between nodes. For each node i, the setup server computes a polynomial share $f(i, y)$, which is derived from the polynomial function $f(x, y)$, and installs $f(i, y)$ into node i prior to deployment. Based on the property of bivariate polynomial, two nodes i and j are able to establish a secure connection if they can compute the same key $f(i, j) = f(j, i)$. This polynomial pool-based scheme is based on the basic polynomial-based key pre-distribution proposed by Blundo *et al.* [Blundo et al., 1992]. Two nodes are able to establish a secure connection only if they share at least one polynomial function in common by exchanging polynomial function IDs. An attacker requires to capture at least $t + 1$ polynomial shares in order to retrieve the original t-degree polynomial function. This scheme shows a significant enhancement of network resilience as long as the number of nodes captured is under a certain threshold, i.e., t in this case. It is interesting to note that this scheme is equivalent to Du *et al.*'s pairwise key pre-distribution scheme discussed above.

Eltoweissy *et al.* [Eltoweissy et al., 2006] developed a protocol for dynamic re-keying in the post-deployment phase. Under the long life cycle assumption, re-keying is necessary in the addition or revocation[4] of nodes. By doing so, node capture attacks can no longer further compromise the rest of the network. When some nodes are suspected to be compromised, the base station sends the re-keying instruction(s) to the cluster controllers[5] to trigger the corresponding re-keying operations. The drawbacks are that their approach requires the coordination among a base station and the cluster controllers. In addition, there is an assumption that the compromised nodes can be detected accurately by the base station.

Based on the design objectives (or constraints consideration) of the above schemes, we summarize this comparison in Table 6.2.

6.10.2 Asymmetric Key-Based Approaches

It is well known that public key cryptosystems are in general more versatile than the symmetric cryptography. They can provide more functions such as digital signature and key exchange. However, they are computationally expensive and undesirable to be implemented in the resource-constrained sensor networks. With advancement of sensor network technologies in recent years, im-

---

[4]Node revocation is used to remove some detected misbehaving nodes.

[5]A cluster controller is responsible for organizing and managing a particular cluster of nodes.

TABLE 6.2: Comparison of popular symmetric trust establishment schemes in WSNs.

Scheme	Energy awareness	Memory consumption	Network resilience	Network connectivity	Additional requirement
Eschenauer [Eschenauer and Gligor, 2002]	Average	High	Poor	Average	None
Jolly [Jolly et al., 2003]	Very good	Very low	Poor	N/A	None
Chan [Chan et al., 2005b]	Good	High	Good	Average	Attack probabilities
Du [Du et al., 2004]	Good	Low	Average	Good	Location information
Liu [Liu et al., 2005a]	Average	High	Good	Average	None
Eltoweissy [Eltoweissy et al., 2006]	Average	Low	Good	Average	None

plementation of asymmetric cryptographic protocols in resource-constrained sensor devices becomes possible. In this section, we review several public key cryptosystems for trust establishment systems in WSNs.

Watro *et al.* [Watro et al., 2004] proposed a set of public key-based protocols, called TinyPK, to support authentication and key agreement between a sensor network and a commonly trusted third party outside the sensor network. They exploited the efficiency of the public operations in RSA [Rivest et al., 1978] cryptosystems with the characteristic that the public operations are very fast compared to other public key technology computations by explicitly choosing some small indices as public keys. Their protocols are specially designed such that the computationally expensive operations are placed on the third parties whenever possible. However, some basic functions, such as revocation of compromised private keys, are not supported. Using their protocols, they demonstrated that the RSA [Rivest et al., 1978] and Diffie-Hellman key agreement techniques [Diffie and Hellman, 1976] can be deployed in existing sensor network devices.

As for energy efficient cryptosystems, there are some other options proposed, such as the XTR public key system [Lenstra and Verheul, 2000], and Elliptic Curve Cryptography (ECC) [Koblitz, 1987, Miller, 1985]. Among them, ECC receives most attention. ECC operates on groups of points over elliptic curves. Its security stems from the hardness of elliptic curve discrete logarithm problem (ECDLP). To solve the integer factorization problem of RSA, there are sub-exponential algorithms. However, only exponential algorithms are known for solving the ECDLP. This is the reason why ECC can achieve the same level of security with smaller key sizes. In the future, it is believed that ECC will dominate the area of developing public key cryptosystems in WSNs [Gura et al., 2004].

Malan *et al.* [Malan et al., 2004] presented a public key infrastructure (PKI) based on ECC to be executed on the MICA2 Mote. They argued that by carefully implementing the multiplication of points on elliptic curves, PKI for secret keys' distribution is tractable on the MICA2 platform. In fact, ECC is believed to offer security computationally equivalent to that of RSA with significantly smaller key size. For instance, a 163-bit ECC key is computationally equivalent to a 768-bit RSA key [Lorincz et al., 2004]. ECC offers an alternative solution to make public key cryptography become feasible on lower power sensor devices.

6.11 Case Study: PPLive

Similar to the incentive aspect discussed in Chapter 5, PPLive does not incorporate any systematic trust management facilities in the client programs. Again this approach works because the PPLive client programs are proprietary and under centralized control. Thus, mutual trust between peers is handled by default. However, when malicious behaviors abound (if they have not, they will), proper trust management controllable at the user level has to be included.

6.12 Summary

Table 6.3 gives a qualitative comparison of different trust management approaches proposed for Internet-based P2P systems. In general, for each peer, local trust scores are based on just a sum of prior transaction ratings. But different approaches are employed for the global aggregation step. Based on the simulation results reported, most of the approaches perform well in terms of convergence rate, trust aggregation accuracy, and message overhead. One particular point to note is that the recently proposed GossipTrust scheme [Zhou and Hwang, 2007a] is the only approach that can work under an unstructured P2P network. This is important because future P2P systems are likely to be unstructured to achieve a higher scalability. Reputation systems [Lethin, 2001] are also considered as promising solutions.

Recently, there are numerous other interesting trust management schemes proposed [Lin et al., 2007, Lu et al., 2007a, Nakajima et al., 2007, Schmidt et al.,

TABLE 6.3: A qualitative comparison of different trust management approaches for P2P systems.

	Local Evaluation	Global Aggregation	Convergence Rate	Trust Accuracy	Message Overhead	Structured Network
Azzedin and Maheswaran [Azzedin and Maheswaran, 2003]	Ratings sum	Combined with recommender's scores	Fast	Moderate	High	Yes
EigenTrust [Kamvar et al., 2003]	Ratings sum	Iterative	Fast	High	Moderate	Yes
PeerTrust [Xiong and Liu, 2004]	Normalized ratings sum	Local computation based on 5 factors	Fast	High	Moderate	Yes
FuzzyTrust [Song et al., 2005]	Fuzzy based	Fuzzy based	Fast	High	Moderate	Yes
SuperTrust [Dimitriou et al., 2007]	Ratings sum	Combined with recommender's scores	Fast	High	High	Yes
PowerTrust [Zhou and Hwang, 2007b]	Bayesian	LRW based	Fast	High	Low	Yes
GossipTrust [Zhou and Hwang, 2007a]	Ratings sum	Gossip based	Fast	High	Low	No

2007, Xu et al., 2007, Zhang and Fang, 2007]. The reader is highly encouraged to study these recent results.

Finally, we have also surveyed in detail practical trust establishment schemes for wireless sensor networks.

6.13   Review Questions

1. What are the essential components of a P2P trust scheme?

2. What are the salient features of the EigenTrust system?

3. How do you compare EigenTrust and PeerTrust?

4. What is the relationship between a trust system and a reputation system?

5. What are the impacts of selfishness in the effectiveness of a trust system?

6. What are the most vulnerable components in a trust system?

7. What is the major challenge in trust establishment in a wireless sensor network?

8. What are the differences in handling trust management between a WSN and an Internet-based P2P system?

Chapter 7

Security Issues

7.1 Overview

Security issues are even more prominent in a P2P network, compared with a traditional client-server system. This is because without centralized authority (i.e., trusted servers), it is very difficult to guarantee data integrity and confidentiality in the P2P data sharing process. Specifically, it is difficult to encrypt data because key management is hard in a P2P network. Without such confidentiality protection, all kinds of serious problems arise, such as file content poisoning, routing table pollution, etc. The key issue here is that there is little a peer can do to verify the data being shared.

Another baffling issue is the use of a P2P network as a vehicle to launch further attacks. The most probable situation is that some malicious peers, by controlling a large number of other peers, can perform DDoS attacks on some specific peers. The way to control a large number of peers can be done through routing table pollution as detailed in this chapter below.

Speaking of controlling the P2P network, it is not necessary to control a large number of physical peers. A malicious peer can actually launch a Sybil attack—to obtain a large number of valid identities on the network. Combating Sybil attacks is therefore a very important area of research. We will discuss some recently proposed schemes below.

Finally, we will also discuss an interesting way to make use of legitimate peers to deliberately poison the contents sent to identified pirates.

7.2 Content Pollution

Pollution attacks refer to the situations where attackers deliberately spread corrupted or faked data in a P2P network. At the very least, the damage is that benign or honest peers' download bandwidth and storage are wasted. In the worst case, such corrupted data could even contain malicious code which can lead to further damages.

Kumar *et al.* [Kumar et al., 2006] presented an interesting mathematical analysis of the proliferation of polluted data in a P2P network. In their model, peers are classified into two types: attackers and benign peers. The former injects polluted data in the network while the latter might inadvertently download them. Here, a complete file is considered as an atomic unit of data. Thus, a downloaded file can either be considered as corrupted and cannot be used, or be considered good as a whole. Furthermore, Kumar *et al.* characterize the peers' behaviors as follows:

1. As soon as a peer downloads a good file, it stops searching for the file.

2. If a peer finds a downloaded file is corrupted, it deletes the file and then searches again. (Kumar *et al.* also considered the case where the peer stops searching after a number of unsuccessful downloads.)

3. After a peer has got a good file, it makes the file available indefinitely in the network. (Kumar *et al.* also considered the case where the peer just leaves the system without contributing the good file, i.e., free-loading.)

4. All peers are homogeneous.

Peers' downloading actions are modeled stochastically. That is, a peer selects a certain version v of a file with a probability, denoted as $q_v(t)$, which is a time-varying quantity. In general, this probability is a function of the availability of different versions and how many copies of each exist at a given time in the network, i.e.,

$$q_v(t) = f_v(n_u(t), u \in V(t)), v \in V(t) \tag{7.1}$$

where $n_u(t)$ is the number of copies of version u, $V(t)$ is the set of different versions of the file, and $f_v(.)$ is a function such that $\sum_{v \in V(t)} q_v(t) = 1$.

Under this framework, Kumar *et al.* considered two different downloading behaviors:

Copy Centric Downloading. In this situation, a peer just randomly chooses a certain copy of the file to download, without regard to its version. Thus, the probability function can be expressed as:

$$q_v(t) = \frac{n_v(t)}{\sum_{u \in V(t)} n_u(t)} \tag{7.2}$$

Version Centric Downloading. In this situation, a peer is sensitive to different versions that exist in the network, and chooses a particular version at random to download. Thus, the probability function can be expressed as:

$$q_v(t) = \frac{1}{|V(t)|} \tag{7.3}$$

Let us consider the Copy Centric Downloading case in a bit more detail. Suppose the attackers inject N polluted copies of the file in the network. Furthermore, suppose that there are M (assumed to be a constant throughout) benign peers in the network and each peer's time spent to inspect a downloaded file (to see if it is a good copy) is exponentially distributed with rate m. Now, we can use x and y to denote the number of benign peers with good and corrupted copies, respectively. Using the tuple (x, y) as a system state, Figure 7.1 depicts the state transitions of the system. Specifically, starting at state (x, y), the system can change to one of the following states:

- $(x + 1, y)$: a peer with no copy gets a good copy;

- $(x, y + 1)$: a peer with no copy downloads a corrupted copy;

- $(x + 1, y - 1)$: a peer with a corrupted copy gets a good copy; and

- (x, y): a peer with a corrupted copy downloads a polluted copy again.

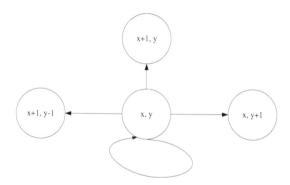

FIGURE 7.1: System state transitions in the Copy Centric Downloading model [Kumar et al., 2006].

Unfortunately, this Markov process is intractable to solve because M is typically a very large number (e.g., 100,000 or more). Thus, Kumar *et al.* [Kumar et al., 2006] resorted to tackling the state transition modeling from an individual peer's perspective. Specifically, at time t, the probability that a peer chooses a corrupted file to download is given by:

$$p(t) = \frac{y(t) + N}{x(t) + y(t) + N} \tag{7.4}$$

A useful insight is that the number of good copies, i.e., $x(t)$, increases when a peer with no copy gets a good copy. This event occurs at the following rate:

$$[M - x(t) - y(t)]m(1 - p(t))$$

Alternatively, $x(t)$ also increases when a peer with a corrupted copy gets a good copy. This event occurs at the following rate:

$$y(t)m(1 - p(t))$$

Combining these two cases, we can express the rate of change of $x(t)$ as follows:

$$\frac{dx(t)}{dt} = [M - x(t) - y(t)]m(1 - p(t)) + y(t)m(1 - p(t)) \qquad (7.5)$$

Similarly, we can also express the rate of change of $y(t)$ as follows:

$$\frac{dy(t)}{dt} = [M - x(t) - y(t)]mp(t) - y(t)m(1 - p(t)) \qquad (7.6)$$

We can also visualize these observations from the state transition diagram shown in Figure 7.2. With some simple algebraic manipulations, the two differential equations can be rewritten as:

$$\frac{dx(t)}{dt} = [M - x(t)]m(1 - p(t)) \qquad (7.7)$$

$$\frac{dy(t)}{dt} = [M - x(t)]mp(t) - my(t) \qquad (7.8)$$

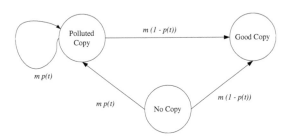

FIGURE 7.2: State transitions for a peer in getting a file in the Copy Centric Downloading model [Kumar et al., 2006].

Kumar *et al.* [Kumar et al., 2006] showed that there is a unique solution:

$$x(t) = \frac{c_2 M (e^{mt} - \frac{c_1}{M+N})^{\frac{M}{M+N}}}{1 + c_2 (e^{mt} - \frac{c_1}{M+N})^{\frac{M}{M+N}}} \qquad (7.9)$$

$$y(t) = M - c_1 e^{-mt} - x(t) \qquad (7.10)$$

where $c_1 = M - x(0) - y(0)$ and

$$c_2 = \frac{x(0)}{M - x(0)} \left(\frac{N + x(0) + y(0)}{M + N} \right)^{-\frac{M}{M+N}}$$

As illustrated by Kumar *et al.*'s numerical results, the above solution generates very accurate data as compared to a simulated system, especially when M is large (e.g., 100,000). Using this model, many interesting conclusions can be drawn. For instance, the number of corrupted copies reaches a peak and then quickly drops off to zero in an exponential manner. Similar mathematical analysis is also given by Kumar *et al.* [Kumar et al., 2006] for the Version Centric Downloading model.

Yang *et al.* [Yang et al., 2008] also proposed a content pollution dynamics model for live video streaming systems. They found that the most critical factors are the access bandwidth and the degree of participating peers, but not the number of initial polluters.

To combat pollution attacks, a typical approach is still based on reputation scores associated with the files as well as the peers that are sharing them. Specifically, when a peer selects a certain file for downloading, it first requests for a weighted sum of votes for the file from a set of peers. The latter sends the requesting peer a vote (e.g., $+1$ or -1) so that a weighted sum can be computed. The weights used represent (or are derived from) the reputations of these responding peers. The systems Credence and Scrubber as described in [Costa and Almeida, 2007] are based on such ideas. A major performance criterion for these systems is the convergence rate. As shown in the simulation results in [Costa and Almeida, 2007], it typically takes several days for the systems to converge to accurately identify true copies from polluted copies.

7.3 Buffer Map Cheating

Li *et al.* [Li et al., 2009] studied the problem of buffer map cheating in P2P video streaming systems. Specifically, they considered the situation where some selfish peers lie about their buffer map contents in that some available chunks are held back. The rationale of this selfish behavior is that the uploading burden can be reduced. However, the streaming quality of the requesting peers could also be reduced.

To combat such malicious behaviors, Li *et al.* [Li et al., 2009] proposed a simple incentive scheme, which works by sorting the requests at a chunk uploading peer in descending order of previous contribution levels. Consequently, if a selfish peer holds back some chunks, its contribution level would be reduced, and thus, it will be placed at a later position in the request queue when it requests chunks from another peer. Li *et al.*'s simulation results show that the proposed simple incentive scheme works quite well in deterring selfish behaviors.

7.4 Sybil Attacks

It is conceivable that a malicious user in the P2P network controls a large number of peers. Specifically, for instance, such a user runs a large number of P2P client programs on the network so that different peers are actually representing the same physical user behind the scene. Such dominance in participation could possibly lead to controlling the operations of the P2P system. For example, the malicious user can easily out-vote other honest peers in the system. This is commonly known as the Sybil attack [Douceur, 2002, Dinger and Hartenstein, 2006]. The root of this problem is the mapping of peer identifiers to physical users.

Dinger and Hartenstein [Dinger and Hartenstein, 2006] formalized the Sybil attack as follows. The set of participants in the P2P network is denoted by: $N = \{p_1, \ldots, p_n\}$. The set of physical users in the network is denoted by: $U = \{u_1, \ldots, u_m\}$. We have $n \geq m$. The set of peers controlled by (or representing) a physical user u_i is denoted by N_i. It follows that $N_i \bigcap N_j$ for all $i \neq j$, and $N_1 \bigcup \ldots \bigcup N_m = N$. Now we can characterize a Sybil attack launched by user u_i as $|N_i| > c$, for a certain constant threshold c.

To prevent Sybil attack, obviously we need a mechanism to avoid assigning multiple peer identifiers to the same physical user. However, under the assumption that a powerful malicious user can modify the P2P client program that is being used, such identifier assignment function cannot be incorporated in the P2P client program. In other words, the assignment process has to be "external" to the client. Now, we have two choices. First, we can use a centralized entity such as a server (e.g., a particular bootstrap server or a tracker) for securely assigning identifiers based on each P2P client's physical attributes (e.g., IP address, MAC address, etc.). However, doing this is somewhat contradictory to the original intent of a P2P network because a centralized authority is involved. Furthermore, this centralized authority would become a single-point-of-failure or single-point-of-attack.

To use a distributed approach in identifier assignment, multiple currently participating peers have to be involved. For example, in the DHT-based algorithm suggested by Dinger and Hartenstein [Dinger and Hartenstein, 2006], multiple peers on a Chord ring have to verify and approve the identifier assigned to a new peer. The identifier is generated based on the IP address of the new peer. A possible loop-hole in this approach is that a malicious user can still get multiple identifiers by using spoofed IP addresses.

Rowaihy et al. [Rowaihy et al., 2007] described another fully distributed approach for identifier assignment in a P2P network. Specifically, their approach is designed for a tree-structured network. In order to be admitted into the network (i.e., obtaining a legal ID), a new peer has to contact a tree-leave peer in the tree. The new peer then needs to solve a cryptographic puzzle which is designed to be resource (e.g., time or memory) constrained. Solving the puzzle is the major barrier in obtaining the admission. The puzzle is cryptographically safe in the sense that it is based on public key cryptography and one-way hash function. Using such an approach, Sybil attack is not eliminated but just becomes much more resource consuming to launch. Thus, if a malicious user is equipped with powerful machines or plenty of resources, a large scale Sybil attack is still possible.

Bazzi and Konjevod [Bazzi and Konjevod, 2005] suggested the use of network coordinates (mentioned in Chapter 4) for honest peers to decide whether new peers are indeed representing distinct physical users (i.e., having very different network coordinates). However, again if a malicious user can manage to spoof multiple physical addresses which are then translated into different network coordinates, the defense is broken.

Yu *et al.* [Yu et al., 2008] proposed a scheme called SybilGuard, relying on topological properties of the social network connecting the physical users. Specifically, based on the social network graph, peers can be logically partitioned into two sets: honest peers and Sybil peers, as shown in Figure 7.3. Here, the links connecting different nodes representing the declared human trust relationships. This topology is independent of and could be entirely different from the P2P network topology. Yu *et al.*'s key insight is that if the malicious user, controlling a large number of peers, can manifest as a very strange structure in the social network graph. Specifically, the graph would have a small *quotient cut*, i.e., a small set of edges (the attack edges) whose removal would partition the graph. As observed empirically, a real-life social network comprising of honest users does not exhibit such a strange property. Thus, a straightforward method to detect Sybil attack is to search for a quotient cut, or to solve the Minimum Quotient Cut problem. Unfortunately this problem is known to be NP-hard.

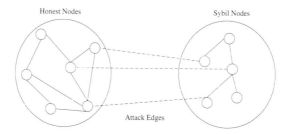

FIGURE 7.3: Based on the social network graph, peers can be classified into honest nodes and Sybil nodes [Yu et al., 2008].

To get around the intractability of the Minimum Quotient Cut problem, Yu *et al.* proposed to use verifiable random walk in the social network graph and the intersections between such walks. Specifically, these random walks are designed so that the small quotient cut (i.e., the set of attack edges) are exploited to defend against the malicious user. Consequently, the number of

Sybil identities is bounded. The random walk scheme is outlined as follows. A peer with degree d carries out d random walks starting from itself of a pre-defined length w (according to Yu *et al.*, $w = 2000$ for a one-million node social network). Now, an honest peer accepts the joining of a new peer if the random walk of the former intersects with the one of the latter. It is not difficult to choose an appropriate w based on the size of the social network so that the random walks of an honest peer reside entirely within the honest peer set. On the other hand, for a malicious new peer's random walk to intersect with some random walks of an honest peer, one of the attack edges must be used. By monitoring the intersection points (which are the incident nodes of edge attack edges), a honest peer can judge probabilistically whether a new peer is indeed a Sybil peer.

7.5   DDoS Attacks

P2P systems are highly vulnerable to be used as a vehicle to launch distributed denial-of-service (DDoS) attack [Naoumov and Ross, 2006]. Specifically, there are two types of attack strategies for exploiting a P2P network: (1) index poisoning; and (2) routing table poisoning.

Every P2P system maintains a certain mapping of keys to values. The most notable example is the mapping of file names to file data. More precisely, file names are eventually mapped to locations of file data. In index poisoning attack, the adversary modifies the index table of peers so that keys are largely mapped to the address of a victim peer, which in fact does not store the requested data. Thus, a swarm of peers requesting certain popular files will make connections with the victim peer, thereby overwhelming it.

In a similar vein, the adversary can also poison the routing table entries of a large number of peers so that routing requests are directed to a victim peer.

Brinkmeier *et al.* [Brinkmeier et al., 2009] also proposed several heuristic techniques to make a P2P live media streaming system more attack resilient. The first insight is that each node should have a low dependency (in terms of data transmission) on other nodes. The rationale is that the streaming quality will not be too affected by node dynamics. The second trick is to balance the relevance and importance of nodes across the network so that the whole system will not be too dependent upon a few key nodes. Finally, it is also important to keep the topology information a secret so that potential malicious peers cannot easily identify important target peers for launching the attack.

7.6 P2P Worm Propagation

Worm propagation over a P2P network has been considered as a highly damaging threat [Xie and Zhu, 2007]. The reason is that the spreading of worms over a P2P network is based on a topological approach—once a certain peer is compromised by a worm, its routing table information can be exploited to specifically target active neighboring peers, without relying on the "traditional" scanning approach. Xie and Zhu [Xie and Zhu, 2007] proposed a heuristic method to combat worm propagation in a P2P network. First, they proposed to proactively select a set of immune peers for blocking the worm. The selection can be based on different schemes. Xie and Zhu considered a partition-based scheme in which the immune peers are chosen in a way that they partition the overlay graph into many nearly balanced sub-graphs. The immune peers in each sub-graph are responsible for blocking the worm in their own regions. They also considered a Connected Dominating Set-(CDS-) based approach in which a security patch is sent to a set of un-infected peers. The set of peers form a dominating set in that every peer not in the subset is adjacent to at least one peer in the subset. A crux in these two schemes is that some "security servers," external to the P2P network, have to carry out the selection task and the security patch delivery.

From another perspective, researchers also consider using a P2P network to combat an outbreak of Internet worms. For instance, Shakkottai and Srikant [Shakkottai and Srikant, 2007] investigated the fundamental insight about propagation of worms under active defense by a P2P network. They derived expressions of orders of magnitude of parameters, such as worm propagation time, maximum number of infected hosts, and patching time. The expressions are based on the following parameters:

- N: total number of hosts in the system;

- β: the maximum rate at which the worm can spread, known as the virulence of the worm, expressed as the number of infections per unit time;

- γ: the ratio of the maximum rate of patch propagation to worm's virulence;

- I_N: the number of infected hosts when the patch is released; and

- P_N: the number of dedicated patch servers.

For example, for a worm like Code-Red, the susceptible population is about 360,000 hosts and the number of infections per hour $\beta = 1.8$ [Staniford et al., 2002].

Shakkottai and Srikant [Shakkottai and Srikant, 2007] showed that with a fixed number of patch servers, the maximum number of infected hosts is

$\Theta(N)$ and the time taken to disinfect the system is $\Theta(\frac{N-2P_N}{P_N})$. For example, for Code-Red worm, roughly 200,000 hosts will be infected in 7 hours and it takes about 25 hours to clean up the system.

On the other hand, with the help of a P2P system, the maximum number of infected hosts is $\Theta(I_N(\frac{N}{P_N})^{\frac{1}{\gamma}})$ and the time taken to disinfect the system is $\Theta(\ln\frac{N}{P_N} + \frac{\gamma}{1+\gamma}\ln I_N)$. For example, for Code-Red worm, even with $\gamma = 2$, the maximum number of infected hosts is on the order of 1000 only and it takes only 5 hours to clean up the system.

7.7   P2P SIP

A key component in VoIP systems, the SIP (Session Initiation Protocol) is also considered as an ideal candidate to be carried by a P2P network, from the robustness perspective [Chopra et al., 2009, Seedorf, 2006]. Specifically, instead of relying on centralized servers, a DHT is employed for registering and locating a user ID (i.e., the SIP-URI). However, while the robustness advantage of this approach is attractive, that is also associated with a whole lot of security concerns. Indeed, most of the security problems that we discuss above could render the P2P SIP protocol crippled. For example, the routing table poisoning attack could make a peer unreachable or overwhelmed with unnecessary traffic.

7.8   Collusive Piracy

Content piracy has always been a serious problem, even before the advent of the Internet. The proliferation use of P2P computing clients only makes this problem worse—many P2P users simply ignore copyright issues to share commercial contents, reducing the potential profits of online content delivery.

Lou and Hwang [Lou and Hwang, 2009] proposed a very interesting approach to combat collusive piracy in a P2P network. Specifically, while content poisoning, as described above, is considered as a vice rather than a virtue, Lou and Hwang suggested a scheme to deliberately poison the paid contents when the system detects that some pirates are downloading them.

In their proposed scheme, each peer is identified with its endpoint address, consisting of an IP address and a listening port number. Notice that for peers behind an NAT device, the public address representing the NAT box is used as the IP address. Files to be shared are then incorporated with digital signatures derived from the endpoint addresses. Legitimate clients can verify the digital signatures based on tokens obtained from the original content source from which the contents are purchased. These tokens are time-stamped and have a short valid time periods. Legitimate clients can periodically refresh the tokens to continue the sharing, while pirates cannot do the same. Upon receiving downloading requests, legitimate clients verify the identities of the requesting peers. If the requesting peers are also legitimate, the digital signatures match those associated with the shared file chunks. On the other hand, pirates are detected because the signatures do not match.

Once a pirate's downloading request is detected, legitimate clients send poisoned chunks to the pirate. Consequently, pirates obtain some poisoned chunks, which will further affect other colluded pirates. The ultimate result is that the download time for a complete file will be too long to be tolerated by the pirates.

7.9 Case Study: PPLive

In many practical P2P systems, sophisticated security measures including chunk encryption, peer authentication, hashes or chunk signatures, etc. are largely not implemented. For instance, in PPLive, peers are essentially unprotected from easy attacks such as pollution attacks, making the user machines highly vulnerable to malicious assaults (e.g., malware spreading). Yet it is very difficult to implement secure and trustworthy data transfer in live video streaming due to practical problems like key distribution and management.

7.10 Summary

P2P security subsumes Internet security. Yet due to its remarkable proliferation, a P2P system's security problems have far more serious adverse consequences. Indeed, one can easily conceive a "nightmare" scenario—a large P2P network is compromised by content poisoning where file contents have malicious codes embedded inside, and then the large number of peers are controlled so as to launch a DDoS attack, at unprecedented scale, to some well-known commercial servers (e.g., eBay). In this scenario, each of the security problems we discussed in this chapter plays a role. Thus, as is always said, P2P security is a system issue so that it is only as strong as its weakest link. Unless we can tackle all the problems satisfactorily, the envisioned nightmare might happen at any time.

7.11   Review Questions

1. Explain content poisoning attack.

2. What is Sybil attack? Why is it a serious problem?

3. Describe a possible scenario of buffer map cheating.

4. Why is it possible to launch a DDoS attack using a P2P network?

5. Why does a P2P network facilitate worm propagation?

Chapter 8

Conclusions

8.1 Where Are We Now?

P2P applications have been proliferating at an ever increasing rate. To many people, especially the younger generations, using some kind of P2P application is already an essential part of daily life. For instance, people routinely visit some P2P web sites when they want to download some files (e.g., CD images, etc.). Furthermore, P2P media applications such as Skype and PPLive are also the default choices for many people. With the advent of smart-phones, it is widely envisioned that such P2P applications will have an even higher penetration as people would very likely use them on the go. Indeed, such development trends are major motivating factors for even large corporations to consider a P2P contents delivery model.

By and large, P2P architectures can still be generally classified into structured and unstructured types. As we have discussed in Chapter 3, both types have their merits and drawbacks. Yet the current trend is that unstructured architectures play a more dominant role because of its inherent robustness and higher scalability. Structured architecture, on the other hand, is mainly used as an auxiliary component for meta-data indexing, e.g., locating a particular tracker server. This trend will likely continue as P2P applications are increasingly used in a wireless and mobile manner.

As discussed in Chapter 4, in order to maintain a specific data sharing architecture, some pre-defined topology control actions have to be taken by participating peers. This is all the more important for structured architectures, which, indeed, have stringent requirements for participants. However, in view of the possibly vigorous peer dynamics, demanding topology control actions might be a nuisance for the peers. This is one of the reasons why an unstructured architecture is more appealing in a large scale P2P system with dynamically changing peer population (i.e., churn). This is fundamentally related to the autonomous and rationally selfish nature of a peer.

A P2P system, at a fundamental level, is nothing more than a dynamic "organism" constituted by a large number of self-optimizing peers. As such, a P2P system's "survival" (in a holistic sense) largely depends on the cooperation among peers. Yet, each peer, being rationally selfish, contributes to the community by "accident" rather than "on purpose." Thus, to ensure that the

P2P system is sustainable, which requires a certain minimum level of cooperation, some incentive schemes have to be in place. That is why, for example, BitTorrent, being the currently most popular file sharing scheme, also has a simple yet effective incentive mechanism (i.e., the tit-for-tat strategy). However, for many other existing P2P systems, there is generally a lack of proper incentive schemes, despite that there are many approaches proposed in the research community.

Similarly, despite that many trust management schemes have been suggested by researchers, they still do not find their way into real life implementations. Trust among peers is still largely based on the "real life trust" among the physical users behind the peers. That is, if a user has interacted with another user for quite some time, there will be a certain level of trust established between them, and thus, such trust is also manifested between the two peers in the P2P network. Autonomous trust management by the peer client programs is still far from a reality. This is true even for some seemingly successful P2P information sharing (or sometimes called static crowd sourcing) applications such as Wikipedia [Wikipedia, 2011] in that the "trust" associated with each information item shared is largely based on offline peer reviewing on the part of the human users.

With a lack of systematic and autonomic trust management, security issues are even more important because a peer might be sharing data with another peer with distrust, which may even be malicious. However, simply put, security is by and large a void in existing P2P systems. People could easily get bogus file data which might even contain some malware or virus. Similar to the situation of trust, security is still based on physical user's judgment—the user manually decides which "torrent" to join to download a file, manually observes whether some links would lead to some phishing sites, etc.

Even when we consider a highly successful P2P streaming system like PPLive in Chapters 5 to 7, there are voids in key aspects such as incentive, trust, and security. Thus, these areas are fertile grounds for high impact research.

8.2   Peer into the Future

"Peering" into the future, we would see P2P applications not just allowing us to share information and processing power but also some more fundamental resources, such as energy in our mobile devices, network bandwidth, security keys, etc. In terms of sharing, different P2P applications could be merged in the sense that getting some video data from a peer would mean giving some file data to it as exchange. That is, in the future, the currently more or less homogeneous "trade" (or barter) would become a wholesale "economy" in which peers trade different kinds of "goods." To make this happen, we would probably need some "medium of exchange," also known as "money" in our physical world. However, such cyberspace "currency" could be totally independent of and has nothing to do with our physical monetary system. Much research needs to be done to realize this futuristic scenario. We do not expect this to be far, though.

With a P2P economy in place, the architecture of the system would be more robust because some peers would be "selling" infrastructure support (i.e., forwarding bandwidth) so that a structured architecture could be revived in many localized regions in the system, mimicking some "cities" in a human being society. Thus, in the future, peers would form clusters and intra-cluster topology would be highly structured, while inter-cluster topology would still be unstructured or in the form of mesh.

Topology control, therefore, becomes a product of some tradings among peers also. Specifically, some peers would be selling infrastructure support to get file data in return (or through some medium of exchange in the process).

Incentive issues then are solved implicitly by the "market" forces, i.e., by demand and supply. Indeed, a peer will be enticed to cooperate if it is given some "commodities" in need for its service. On the other hand, some "services" in peers would be eliminated by the market if there is little or even no demand for them.

Autonomous trust among peers would still be a difficult problem to solve. Perhaps this is because trust is really a hard-to-quantify notion. A possible implementation could be based on fuzzy logic. Yet even if we use fuzzy logic, we cannot do away with the inherent stochastic nature of trust, making it possible to have erroneous actions.

As mentioned at the beginning of this section, even security resources such as keys could be commodities for trade. Thus, security problems could also be handled by the "market."

Finally, a curious question is whether some form of "government" would emerge. If so, its manifestation in a P2P system would defy the original purpose of autonomous sharing/trading.

Bibliography

[Aberer, 2001] Aberer, K. (2001). P-Grid: A Self-Organizing Access Structure for P2P Information Systems. In *Proceedings of the 9th International Conference on Cooperative Information Systems*.

[Abrams et al., 2005] Abrams, Z., McGrew, R., and Plotkin, S. (2005). A Non-Manipulable Trust System Based on EigenTrust. *ACM SIGecom Exchanges*, (4):21–30.

[Ahlswede et al., 2000] Ahlswede, R., Cai, N., Li, S.-Y. R., and Yeung, R. W. (2000). Network Information Flow. *IEEE Transactions on Information Theory*, (7):1204–1216.

[Akyildiz et al., 2002] Akyildiz, I. F., Weilian, S., Sankarasubramaniam, Y., and Cayirci, E. (2002). A survey on sensor networks. *IEEE Communications Magazine*, 40(8):102–114.

[Akyol et al., 2006] Akyol, E., Tekalp, A. M., and Civanlar, R. R. (2006). Adaptive Peer-to-Peer Video Streaming with Optimized Flexible Multiple Description Coding. In *Proceedings of the 2006 International Conference on Image Processing*, pages 725–728.

[Anagnostakis and Greenwald, 2004] Anagnostakis, K. G. and Greenwald, M. B. (2004). Exchange-Based Incentive Mechanisms for Peer-to-Peer File Sharing. In *Proceedings of the 24th International Conference on Distributed Computing Systems*.

[Anderson et al., 2002] Anderson, D. P., Cobb, J., Korpela, E., Lebofsky, M., and Werthimer, D. (2002). SETIhome: An Experminent in Public-Resource Computing. *Communications of the ACM*, (11):56–61.

[Anderson et al., 2004] Anderson, R., Chan, H., and Perrig, A. (2004). Key Infection: Smart Trust for Smart Dust. In *ICNP 2004*, pages 206–215.

[Androutsellis-Theotokis and Spinellis, 2004] Androutsellis-Theotokis, S. and Spinellis, D. (2004). A Survey of Peer-to-Peer Content Distribution Technologies. *ACM Computing Surveys*, 36(4):335–371.

[Auvinen et al., 2007] Auvinen, A., Vapa, M., Weber, M., Kotilainen, N., and Vuori, J. (2007). New Toplogy Management Algorithms for Unstructured

P2P Networks. In *Proceedings of the 2nd International Conference on Internet and Web Applications and Services (ICIW)*.

[Azzedin and Maheswaran, 2003] Azzedin, F. and Maheswaran, M. (2003). Trust Modeling for Peer-to-Peer Based Computing Systems. In *Proceedings 17th International Parallel and Distributed Symposium (IPDPS)*.

[Azzedin and Maheswaran, 2004] Azzedin, F. and Maheswaran, M. (2004). A Trust Brokering System and Its Application to Resource Management in Public-Resource Grids. In *Proceedings 18th International Parallel and Distributed Symposium (IPDPS)*.

[Balakrishnan et al., 2003] Balakrishnan, H., Kaashoek, M. F., Karger, D., Morris, R., and Stoica, I. (2003). Looking Up Data in P2P Systems. *Communications of the ACM*, (2):43–48.

[Barabasi and Albert, 1999] Barabasi, A.-L. and Albert, R. (1999). Emergence of Scaling in Random Networks. *Science*, (5439):509–512.

[Baset and Schulzrinne, 2006] Baset, S. A. and Schulzrinne, H. G. (2006). An Analysis of the Skype Peer-to-Peer Internet Telephony Protoco. In *Proceedings of INFOCOM 2006*.

[Bazzi and Konjevod, 2005] Bazzi, R. and Konjevod, G. (2005). On the Establishment of Distinct Identities in Overlay Networks. In *Proceedings of the 24th ACM Symposium on Principles of Distributed Computing (PODC 2005)*, pages 312–320.

[Becker and Clement, 2004] Becker, J. U. and Clement, M. (2004). The Economic Rationale of Offering Media Files in Peer-to-Peer Networks. In *Proceedings of the 37th Hawaii International Conference on System Sciences*.

[Bertino et al., 2004] Bertino, E., Ferrari, E., and Squicciarini, A. C. (2004). Trust-χ: A Peer-to-Peer Framework for Trust Establishment. *IEEE Transactions on Knowledge and Data Engineering*, (7):827–842.

[Bharghavan et al., 1994] Bharghavan, V., Demers, A., Shenker, S., and Zhang, L. (1994). MACAW: A Media Access Protocol for Wireless LAN's. In *ACM SIGCOMM Computer Communication Review*, pages 212–225.

[BitComet, 2009] BitComet (2009). http://www.bitcomet.com.

[BitTorrent, 2009] BitTorrent (2009). http://www.bittorrent.com.

[Blom, 1984] Blom, R. (1984). An Optimal Class of Symmetric Key Generation Systems. In *EUROCRYPT Workshop on Advances in Cryptology*, pages 335–338.

[Blundo et al., 1992] Blundo, C., Santis, A. D., Herzberg, A., Kutten, S., Vaccaro, U., and Yung, M. (1992). Perfectly Secure Key Distribution for Dynamic Conferences. In *CRYPTO 1992*, volume 740, pages 471–486.

[BOINC, 2009] BOINC (2009). http://boinc.berkeley.edu.

[Bonald et al., 2008] Bonald, T., Massoulie, L., Mathieu, F., Perino, D., and Twigg, A. (2008). Epidemic Live Streaming: Optimal Performance Trade-Offs. In *Proceedings of ACM SIGMETRICS 2008*, pages 325–336.

[Brinkmeier et al., 2009] Brinkmeier, M., Schafer, G., and Strufe, T. (2009). Optimally DoS Resistant P2P Topologies for Live Multimedia Streaming. *IEEE Transactions on Parallel and Distributed Systems*, (6):831–844.

[Buchegger and Boudec, 2005] Buchegger, S. and Boudec, J.-Y. L. (2005). Self-Policing Mobile Ad Hoc Networks by Reputation Systems. *IEEE Communications Magazine*, pages 101–107.

[Butler et al., 2000] Butler, R., Welch, V., Engert, D., Foster, I., Tuecke, S., Volmer, J., and Kesselman, C. (2000). A National-Scale Authentication Infrastructure. *IEEE Computer*, (12):60–66.

[Caizzone et al., 2008] Caizzone, G., Corghi, A., Giacomazzi, P., and Nonnoi, M. (2008). Analysis of the Scalability of the Overlay Skype System. In *Proceedings of ICC 2008*.

[Castro et al., 2003a] Castro, M., Druschel, P., Kermarrec, A., Nandi, A., Rowstron, A., and Singh, A. (2003a). Splitstream: High-Bandwidth Multicast in Cooperative Environments. In *Proceedings of the 19th ACM Symposium on Operating Systems Principles*.

[Castro et al., 2003b] Castro, M., Druschel, P., Kermarrec, A.-M., Nandi, A., Rowstron, A., and Singh, A. (2003b). SplitStream: High-Bandwidth Multicast in Cooperative Environments. In *Proceedings of SOSP 2003*, pages 298–313.

[Chan et al., 2005a] Chan, H., Gligor, V. D., Perrig, A., and Muralidharan, G. (2005a). On the Distribution and Revocation of Cryptographic Keys in Sensor Networks. *IEEE Transactions on Dependable and Secure Computing*, 2(3):233–247.

[Chan and Perrig, 2005] Chan, H. and Perrig, A. (2005). PIKE: Peer Intermediaries for Key Establishment in Sensor Networks. In *INFOCOM 2005*, volume 1, pages 524–535.

[Chan et al., 2003] Chan, H., Perrig, A., and Song, D. (2003). Random Key Predistribution Schemes for Sensor Networks. In *IEEE Symposium on Security and Privacy*, pages 197–213.

[Chan et al., 2005b] Chan, S.-P., Poovendran, R., and Sun, M.-T. (2005b). A Key Management Scheme in Distributed Sensor Networks Using Attack Probabilities. In *GLOBECOM 2005*, volume 2, pages 1007–1011.

[Chopra et al., 2009] Chopra, D., Schulzrinne, H., Marocco, E., and Ivov, E. (2009). Peer-to-Peer Overlays for Real-Time Communication: Security Issues and Solutions. *IEEE Communications Survey and Tutorials*, (1):4–12.

[Cisco, 2009] Cisco (2009). Technical Specification of Cisco AIR-CB21AG. http://www.cisco.com.

[Clarke et al., 2000] Clarke, I., Sandberg, O., Wiley, B., and Hong, T. (2000). Freenet: A Distributed Anonymous Information Storage and Retrieval System. In *Proceedings of the Workshop on Design Issues in Anonymous and Unobservability*.

[Cohen, 2003] Cohen, B. (2003). Incentives Build Robustness in BitTorrent. In *Proceedings of the Workshop on Economics of Peer-to-Peer Systems*.

[CoolStreaming, 2009] CoolStreaming (2009). http://webtv.coolstreaming.us.

[Costa and Almeida, 2007] Costa, C. and Almeida, J. (2007). Reputation Systems for Fighting Pollution in Peer-to-Peer File Sharing Systems. In *Proceedings of the Seventh IEEE International Conference on Peer-to-Peer Computing*.

[Courcoubetis and Weber, 2006] Courcoubetis, C. and Weber, R. (2006). Incentives for Large Peer-to-Peer Systems. *IEEE Journal on Selected Areas in Communications*, 24(5):1034–1050.

[Crossbox Technology, 2008] Crossbox Technology (2008). http://www.xbow.com/.

[Cui and Nahrstedt, 2003] Cui, Y. and Nahrstedt, K. (2003). Layered Peer-to-Peer Streaming. In *Proceedings of NOSSDAV'03*, pages 162–171.

[Dabek et al., 2004] Dabek, F., Cox, R., Kaashoek, F., and Morris, R. (2004). Vivaldi: A Decentralized Network Coordinate System. In *Proceedings of ACM SIGCOMM*.

[Dale et al., 2008] Dale, C., Liu, J., Peters, J., and Li, B. (2008). Evolution and Enhancement of BitTorrent Network Topologies. In *Proceedings of IWQoS 2008*.

[Diffie and Hellman, 1976] Diffie, W. and Hellman, M. (1976). New Directions in Cryptography. *IEEE Transactions on Information Theory*, 22:644–654.

[Dimitriou et al., 2007] Dimitriou, T., Karame, G., and Christou, I. (2007). SuperTrust—A Secure and Efficient Framework for Handling Trust in Super-Peer Networks. In *Proceedings PODC 2007*.

[Dinger and Hartenstein, 2006] Dinger, J. and Hartenstein, H. (2006). Defending the Sybil Attack in P2P Networks: Taxonomy, Challenges, and a Proposal for Self-Registration. In *Proceedings of the First International Conference on Availability, Reliability and Security (ARES 2006)*.

[Douceur, 2002] Douceur, J. (2002). The Sybil Attack. In *Proceedings of the 1st International Workshop on Peer-to-Peer Systems (IPTPS 2002)*.

[Du et al., 2004] Du, W., Deng, J., Han, Y. S., Chen, S., and Varshney, P. K. (2004). A Key Management Scheme for Wireless Sensor Networks Using Deployment Knowledge. In *INFOCOM 2004*, volume 1, pages 586–597.

[Du et al., 2005] Du, W., Deng, J., Han, Y. S., Varshney, P. K., Katz, J., and Khalili, A. (2005). A Pairwise Key Predistribution Scheme for Wireless Sensor Networks. *ACM Transactions on Information and System Security*, 8(2):228–258.

[eDonkey, 2009] eDonkey (2009). `http://www.brothersoft.com/downloads/edonkey.html`.

[Efstathiou and Polyzos, 2003] Efstathiou, E. C. and Polyzos, G. C. (2003). A Peer-to-Peer Approach to Wireless LAN Roaming. In *Proceedings of WMASH*, pages 10–18.

[Einstein@Home, 2009] Einstein@Home (2009). `http://einstein.phys.uwm.edu`.

[Eltoweissy et al., 2006] Eltoweissy, M., Moharrum, M., and Mukkamala, R. (2006). Dynamic Key Management in Sensor Networks. *IEEE Communications Magazine*, 44(4):122–130.

[eMule, 2009] eMule (2009). `http://www.emule-project.net`.

[Erdös and Rényi, 1960] Erdös, P. and Rényi, A. (1960). On the Evolution of Random Graph. *Publ. Math. Inst. Hung. Acad. Sci.*, 5:17–61.

[Eschenauer and Gligor, 2002] Eschenauer, L. and Gligor, V. D. (2002). A Key-Management Scheme for Distributed Sensor Networks. In *The 9th ACM Conference on Computer and Communications Security*, pages 41–47.

[Feldman and Chuang, 2005] Feldman, M. and Chuang, J. (2005). Overcoming Free-Riding Behavior in Peer-to-Peer Systems. *ACM SIGccom Exchanges*, 5(4):41–50.

[Feldman et al., 2004a] Feldman, M., Lai, K., Stoica, I., and Chuang, J. (2004a). Robust Incentive Techniques for Peer-to-Peer Networks. In *Proceedings of the 5th ACM conference on Electronic Commerce*, pages 102–111.

[Feldman et al., 2004b] Feldman, M., Papadimitriou, C., Chuang, J., and Stoica, I. (2004b). Free-Riding and Whitewashing in Peer-to-Peer Systems. In *Proceedings of the 2004 SIGCOMM Workshop on Practice and Theory of Incentives in Networked Systems*, pages 228–235.

[Felegyhazi et al., 2006] Felegyhazi, M., Hubaux, J.-P., and Buttyan, L. (2006). Nash Equilibria of Packet Forwarding Strategies in Wireless Ad Hoc Networks. *IEEE Transactions on Mobile Computing*, 5(5):463–476.

[Feng and Li, 2008] Feng, C. and Li, B. (2008). On Large-Scale Peer-to-Peer Streaming Systems with Network Coding. In *Proceedings of ACM Multimedia 2008*, pages 269–278.

[Figueiredo et al., 2005] Figueiredo, D., Shapiro, J., and Towsley, D. (2005). Incentives to Promote Availability in Peer-to-Peer Anonymity Systems. In *Proceedings of the 13th IEEE International Conference on Network Protocols*.

[Folding@Home, 2009] Folding@Home (2009). `http://folding.stanford.edu`.

[Foster and Kesselman, 1999] Foster, I. and Kesselman, C. (1999). *The Grid: Blueprint for a New Computing Infrastructure*. San Francisco.

[Foxy, 2009] Foxy (2009). `http://tw.myfoxy.net`.

[Frey and Murphy, 2008] Frey, D. and Murphy, A. L. (2008). Failure-Tolerant Overlay Trees for Large-Scale Dynamic Networks. In *Proceedings of the Eighth International Conference on Peer-to-Peer Computing (P2P 2008)*, pages 351–361.

[Fu et al., 2005] Fu, H., Kawamura, S., Zhang, M., and Zhang, L. (2005). Replication Attack on Random Key Pre-Distribution Schemes for Wireless Sensor Networks. In *The 6th Annual IEEE SMC Information Assurance Workshop*, pages 134–141.

[Fu et al., 2008] Fu, L., Qu, H., Chen, H., Wang, H., and Wang, X. (2008). A Hierarchical and Heterogeneous P2P-SIP Architecture. In *Proceedings of ICPCA 2008*, pages 995–998.

[Ge et al., 2003] Ge, Z., Figueiredo, D. R., Sharad, J., Kurose, J., and Towsley, D. (2003). Modeling Peer-Peer File Sharing Systems. In *Proceedings IEEE INFOCOM 2003*, pages 2188–2198.

[giFT-FastTrack, 2009] giFT-FastTrack (2009). `http://developer.berlios.de/projects/gift-fasttrack`.

[Gkantsidis and Rodriguez, 2005] Gkantsidis, C. and Rodriguez, P. (2005). Network Coding for Large Scale Content Distribution. In *Proceedings IEEE INFOCOM*.

[Gnutella Protocol Development, 2009] Gnutella   Protocol   Development (2009). http://rfc-gnutella.sourceforge.net.

[Golle et al., 2001] Golle, P., Leyton-Brown, K., and Mironov, I. (2001). Incentives for Sharing in Peer-to-Peer Networks. In *Proceedings of the ACM Conference on Electronic Commerce*, pages 264–267.

[Goyal, 2001] Goyal, V. K. (2001). Multiple Description Coding: Compression Meets the Network. *IEEE Signal Processing Magazine*, (5):74–93.

[GreenTea Technologies Inc., 2009] GreenTea Technologies Inc. (2009). http://www.greenteatech.com.

[Griffiths et al., 2006] Griffiths, N., Chao, K.-M., and Younas, M. (2006). Fuzzy Trust for Peer-to-Peer Systems. In *Proceedings 26th IEEE International Conference on Distributed Computing Systems Workshops*.

[GTRAN, 2009] GTRAN (2009). Technical Specification of GTRAN Dot-Surfer 6210. http://www.gtran.com.

[Gu et al., 2008] Gu, X., Wen, Z., Yu, P. S., and Shae, Z.-Y. (2008). peerTalk: A Peer-to-Peer Multiparty Voice-over-IP System. *IEEE Transactions on Parallel and Distributed Systems*, 19(4):515–528.

[Gupta and Somani, 2004] Gupta, R. and Somani, A. K. (2004). A Pricing Strategy for Incentivizing Selfish Nodes to Share Resources in Peer-to-Peer (P2P) Networks. In *Proceedings of the IEEE International Conference on Networks*.

[Gura et al., 2004] Gura, N., Patel, A., Wander, A., Eberle, H., and Shantz, S. C. (2004). Comparing Elliptic Curve Cryptography and RSA on 8-bit CPUs. In *Workshop on Cryptographic Hardware and Embedded Systems*, pages 119–132.

[Habib and Chuang, 2006] Habib, A. and Chuang, J. (2006). Service Differentiated Peer Selection: An Incentive Mechanism for Peer-to-Peer Media Streaming. *IEEE Transactions on Multimedia*, 8(3):601–621.

[Hariri et al., 2007] Hariri, B., Shirmohammadi, S., and Pakravan, M. R. (2007). A Distributed Toplogy Control Algorithm for P2P Based Simulations. In *Proceedings of the 11th IEEE Symposium on Distributed Simulation and Real-Time Applications*.

[Hastings, 1970] Hastings, W. (1970). Monte Carlo Sampling Methods Using Markov Chains and Their Applications. *Biometrika*, (1):97–109.

[Hausheer et al., 2003] Hausheer, D., Liebau, N. C., Mauthe, A., Steinmetz, R., and Stiller, B. (2003). Token Based Accounting and Distributed Pricing to Introduce Market Mechanisms in a Peer-to-Peer File Sharing Scenario. In *Proceedings of the Third International Conference on Peer-to-Peer Computing*.

[Hausheer and Stiller, 2005] Hausheer, D. and Stiller, B. (2005). Decentralized Auction-Based Pricing with PeerMart. In *Proceedings of the 9th IFIP/IEEE International Symposium on Integrated Network Management*.

[Hefeeda et al., 2003] Hefeeda, M., Habib, A., Botev, B., Xu, D., and Bhargava, B. (2003). PROMISE: Peer-to-Peer Media Streaming Using Collect-Cast. In *Proceedings of the ACM Multimedia*.

[Hei et al., 2007a] Hei, X., Liang, C., Liang, J., Liu, Y., and Ross, K. W. (2007a). A Measurement Study of a Large-Scale P2P IPTV System. *IEEE Transactions on Multimedia*, 9(8):1672–1687.

[Hei et al., 2007b] Hei, X., Liu, Y., and Ross, K. W. (2007b). Inferring Network-Wide Quality in P2P Live Streaming Systems. *IEEE Journal on Selected Areas in Communications*, 25(9):1640–1654.

[Horvath et al., 2008] Horvath, A., Telek, M., Rossi, D., Veglia, P., Ciullo, D., Garcia, M. A., Leonardi, E., and Mellia, M. (2008) Dissecting PPLive, SopCast, TVAnts.

[Hsiao and King, 2003] Hsiao, H.-C. and King, C.-T. (2003). Bristle: A Mobile Structured Peer-to-Peer Architecture. In *Proceedings of IPDPS 2003*.

[Hsieh and Sivakumar, 2004] Hsieh, H.-Y. and Sivakumar, R. (2004). On Using Peer-to-Peer Communication in Cellular Wireless Data Networks. *IEEE Transactions on Mobile Computing*, 3(1):57–72.

[hua Chu and Zhang, 2004] hua Chu, Y. and Zhang, H. (2004). Considering Altruism in Peer-to-Peer Internet Streaming Broadcast. In *Proceedings of NOSSDAV'04*, pages 10–15.

[Huang et al., 2007] Huang, C., Li, J., and Ross, K. W. (2007). Can Internet Video-on-Demand Be Profitable? In *Proceedings of SIGCOMM 2007*, pages 133–144.

[Huang et al., 2008] Huang, Y., Fu, T. Z. J., Chiu, D.-M., Lui, J. C. S., and Huang, C. (2008). Challenges, Design and Analysis of a Large-Scale P2P-VoD System. In *Proceedings of ACM SIGCOMM 2008*, pages 375–388.

[Ileri et al., 2005] Ileri, O., Mau, S.-C., and Mandayam, N. B. (2005). Pricing for Enabling Forwarding in Self-Configuring Ad Hoc Networks. *IEEE Journal on Selected Areas in Communications*, 23(1):151–162.

[iMesh, 2009] iMesh (2009). http://www.imesh.com.

[Jafarisiavoshani et al., 2007] Jafarisiavoshani, M., Fragouli, C., Diggavi, S., and Gkantsidis, C. (2007). Bottleneck Discovery and Overlay Management in Network Coded Peer-to-Peer Systems. In *Proceedings ACM INM*, pages 293–298.

[Jia et al., 2005] Jia, Z., Tiange, S., Liansheng, H., and Yiqi, D. (2005). A New Micropayment Protocol Based on P2P Networks. In *Proceedings of the 2005 IEEE International Conference on e-Business Engineering.*

[Jolly et al., 2003] Jolly, G., Kuscu, M. C., Kokate, P., and Younis, M. (2003). A Low-Energy Key Management Protocol for Wireless Sensor Networks. In *The 8th IEEE International Symposium on Computers and Communication,* volume 1, pages 335–340.

[Joost, 2009] Joost (2009). http://www.joost.com.

[Jun and Ahamad, 2005] Jun, S. and Ahamad, M. (2005). Incentives in Bit-Torrent Induce Free Riding. In *Proceedings of the ACM SIGCOMM 2005 Workshop.*

[Jun et al., 2005] Jun, S., Ahamad, M., and Xu, J. J. (2005). Robust Information Dissemination in Uncooperative Environments. In *Proceedings of the 25th IEEE International Conference on Distributed Computing Systems.*

[Kalogeraki et al., 2003] Kalogeraki, V., Delis, A., and Gunopulos, D. (2003). Peer-to-Peer Architectures for Scalable, Efficient, and Reliable Media Services. In *Proceedings of IPDPS 2003.*

[Kamvar et al., 2003] Kamvar, S. D., Schlosser, M. T., and Garcia-Molina, H. (2003). The EigenTrust Algorithm for Reputation Management in P2P Networks. In *Proceedings WWW 2003.*

[Kang and Mutka, 2005] Kang, S.-S. and Mutka, M. W. (2005). A Mobile Peer-to-Peer Approach for Multimedia Content Sharing Using 3G/WLAN Dual Mode Channels. *Wireless Communications and Mobile Computing,* 5:633–645.

[Karger et al., 1997] Karger, D., Lehman, E., Leighton, F., Levine, M., Lewin, D., and Panigrahy, R. (1997). Consistent Hashing and Random Trees: Distributed Caching Protocols for Relieving Hot Spots on the World Wide Web. In *Proceedings of the 29th Annual ACM Symposium on Theory of Computing,* pages 654–663.

[KaZaA, 2009] KaZaA (2009). http://www.kazaa.com.

[Kho et al., 2008] Kho, W., Baset, S. A., and Schulzrinne, H. G. (2008). Skype Relay Calls: Measuerments and Experiments. In *Proceedings of INFOCOM 2008.*

[Koblitz, 1987] Koblitz, N. (1987). Elliptic Curve Cryptosystems. *Mathematics of Computation,* 48(177):203–209.

[Krishnana et al., 2003] Krishnana, R., Smith, M. D., and Telang, R. (2003). The Economics of Peer-to-Peer Networks. *Journal of Information Technology Theory and Application,* 5(3):31–44.

[Kumar et al., 2006] Kumar, R., Yao, D. D., Bagchi, A., Ross, K. W., and Rubenstein, D. (2006). Fluid Modeling of Pollution Proliferation in P2P Networks. In *Proceedings of the ACM SIGMETRICS*, pages 335–346.

[Kwok, 2007] Kwok, Y.-K. (2007). Key Management in Wireless Sensor Networks. In *Security in Distributed and Networking Systems*, Yang Xiao and Yi Pan (eds.), World Scientific Publishing Co.

[Kwong and Tsang, 2008] Kwong, K.-W. and Tsang, D. H. K. (2008). Building Heterogeneous Peer-to-Peer Networks: Protocol and Analysis. *IEEE/ACM Transactions on Networking*, (2):281–292.

[Lagesse and Kumar, 2007] Lagesse, B. and Kumar, M. (2007). UBCA: Utility-Based Clustering Architecture for Peer-to-Peer Systems. In *Proceedings of ICDCS Workshop 2007*.

[Lenstra and Verheul, 2000] Lenstra, A. K. and Verheul, E. R. (2000). The XTR public key system. In *The 20th Annual International Cryptology Conference on Advances in Cryptology*, pages 1–19.

[Lethin, 2001] Lethin, R. (2001). Reputation. In *Peer-to-Peer: Harnessing the Benefits of a Disruptive Technology*, pages 341–353.

[Leuf, 2002] Leuf, B. (2002). *Peer-to-Peer: Collaboration and Sharing over the Internet*. Boston.

[Leung and Kwok, 2005a] Leung, A. K. H. and Kwok, Y.-K. (2005a). An Efficient and Practical Greedy Algorithm for Server-Peer Selection in Wireless Peer-to-Peer File Sharing Networks. In *Proceedings of the International Conference on Mobile Ad-hoc and Sensor Networks (MSN'2005)*, pages 1016–1025.

[Leung and Kwok, 2005b] Leung, A. K. H. and Kwok, Y.-K. (2005b). Community-Based Asynchronous Wakeup Protocol for Wireless Peer-to-Peer File Sharing Networks. In *Proceedings of the IEEE Second Annual International Conference on Mobile and Ubiquitous Systems: Networking and Services (MobiQuitous'2005)*, pages 342–350.

[Leung and Kwok, 2005c] Leung, A. K. H. and Kwok, Y.-K. (2005c). Energy Conservation by Peer-to-Peer Relaying in Quasi-Ad Hoc Networks. In *Proceedings of the IFIP International Conference on Network and Parallel Computing (NPC'2005)*, pages 451–460.

[Leung and Kwok, 2005d] Leung, A. K. H. and Kwok, Y.-K. (2005d). On Topology Control of Wireless Peer-to-Peer File Sharing Networks: Energy Efficiency, Fairness and Incentive. In *Proceedings of the IEEE International Symposium on a World of Wireless, Mobile and Multimedia Networks (WoWMoM'2005)*, pages 318–323.

[Leung and Kwok, 2008] Leung, A. K.-H. and Kwok, Y.-K. (2008). On Localized Application-Driven Topology Control for Energy-Efficient Wireless Peer-to-Peer File Sharing. *IEEE Transactions on Mobile Computing*, (1):66–80.

[Li et al., 2007] Li, B., Xie, S., Keung, G. Y., Liu, J., Stoica, I., Zhang, H., and Zhang, X. (2007). An Empirical Study of the CoolStreaming+ System. *IEEE Journal on Selected Areas in Communications*, (9):1627–1639.

[Li et al., 2009] Li, D., Wu, J., and Cui, Y. (2009). Defending Against Buffer Map Cheating in DONet-Like P2P Streaming. *IEEE Transactions on Multimedia*, (3):535–542.

[Li and Singhal, 2007] Li, H. and Singhal, M. (2007). Trust Management in Distributed Systems. *IEEE Computer*, pages 45–53.

[Li et al., 2008] Li, Z., Yu, Y., Hei, X., and Tsang, D. H. K. (2008). Towards Low-Redundany Push-Pull P2P Live Streaming. In *Proceedings of QShine 2008*.

[Liang et al., 2005] Liang, J., Kumar, R., and Ross, K. W. (2005). The KaZaA Overlay: A Measurement Study. *Computer Networks*.

[Lin et al., 2007] Lin, Z., Feng, X., Yuan, W., and Jian, L. (2007). A Semantic and Time Related Recommendation-Feedback Trust Model. In *Proceedings 2nd International Conference Availability, Reliability, and Security (ARES)*.

[Liu and Ning, 2003] Liu, D. and Ning, P. (2003). Location-Based Pairwise Key Establishments for Static Sensor Networks. In *The 1st ACM Workshop on Security of Ad Hoc and Sensor Networks*, pages 72–82.

[Liu et al., 2005a] Liu, D., Ning, P., and Li, R. (2005a). Establishing Pairwise Keys in Distributed Sensor Networks. *ACM Transactions on Information and System Security*, 8(1):41–77.

[Liu et al., 2005b] Liu, Y., Xiao, L., Liu, X., Ni, L. M., and Zhang, X. (2005b). Location Awareness in Unstructured Peer-to-Peer Systems. *IEEE Transactions on Parallel and Distributed Systems*, (2):163–174.

[Lorincz et al., 2004] Lorincz, K., Malan, D. J., Fulford-Jones, T. R., Nawoj, A., Clavel, A., Shnayder, V., Mainland, G., Welsh, M., and Moulton, S. (2004). Sensor Networks for Emergency Response: Challenges and Opportunities. *IEEE Pervasive Computing*, 3(4):16–23.

[Lou and Hwang, 2009] Lou, X. and Hwang, K. (2009). Collusive Piracy Prevention in P2P Content Deliver Networks. *IEEE Transactions on Computers*, (7):970–983.

[Lu et al., 2007a] Lu, L., Han, J., Hu, L., Huai, J., Liu, Y., and Ni, L. M.
  (2007a). Pseudo Trust: Zero-Knowledge Based Authentication in Anonymous Peer-to-Peer Protocols. In *Proceedings 22nd International Parallel
  and Distributed Symposium (IPDPS)*.

[Lu et al., 2007b] Lu, M.-T., Wu, J.-C., Peng, K.-J., Huang, P., Yao, J. J.,
  and Chen, H. H. (2007b). Design and Evaluation of a P2P IPTV System
  for Heterogeneous Networks. *IEEE Transactions on Multimedia*, (8):1568–
  1579.

[Lv et al., 2002] Lv, Q., Cao, P., Cohen, E., Li, K., and Shenker, S. (2002).
  Search and Replication in Unstructured Peer-to-Peer Networks. In *Pro-
  ceedings of the 16th ACM International Conference on Supercomputing
  (ICS'02)*, pages 84–95.

[Ma and Zhu, 2008] Ma, L. and Zhu, W. (2008). A Carrier Grade Peer-to-
  Peer Network Architecture. In *Proceedings of the 1st ITU-T Kaleidoscope
  Academic Conference*.

[Ma et al., 2004a] Ma, R. T. B., Lee, S. C. M., Lui, J. C. S., and Yau, D.
  K. Y. (2004). A Game Theoretic Approach to Provide Incentive and Service
  Differentiation in P2P Networks Richard T. B. Ma and Sam C. M. Lee and
  John C. S. Lui and David K. Y. Yau. In *Proceedings of SIGMETRICS*,
  pages 189–198.

[Ma et al., 2004b] Ma, R. T. B., Lee, S. C. M., Lui, J. C. S., and Yau, D.
  K. Y. (2004). An Incentive Mechanism for P2P Networks. In *Proceedings
  of the 24th International Conference on Distributed Computing Systems*.

[Malan et al., 2004] Malan, D. J., Welsh, M., and Smith, M. D. (2004). A
  Public-Key Infrastructure for Key Distribution in Tinyos Based on Elliptic
  Curve Cryptography. In *The 1st Annual IEEE Communications Society
  Conference on Sensor and Ad Hoc Communications and Networks*, pages
  71–80.

[Marbach and Qiu, 2005] Marbach, P. and Qiu, Y. (2005). Cooperation in
  Wireless Ad Hoc Networks: A Market Based Approach. *IEEE/ACM Transactions on Networking*, 13(6):1325–1338.

[Marsh and Dibben, 2005] Marsh, S. and Dibben, M. R. (2005). Trust, Untrust, Distrust, and Mistrust—An Exploration of the Dark(er) Side. In
  Proceedings International Conference Trust Management, pages 17–33.

[Maze, 2006] Maze (2006). http://maze.pku.edu.cn.

[Metropolis et al., 1953] Metropolis, N., Rosenbluth, A., Rosenbluth, M.,
  Teller, A., and Teller, E. (1953). Equation of State Calculations by Fast
  Computing Machines. *The Journal of Chemical Physics*, (6):1087–1092.

[Milgram, 1967] Milgram, S. (1967). The Small World Problem. *Psychology Today*, (1):60–67.

[Miller, 1985] Miller, V. S. (1985). Use of Elliptic Curves in Cryptography. *Lecture Notes in Computer Science: Advances in Cryptology*, 218:417–426.

[Milojicic et al., 2002] Milojicic, D. S., Kalogeraki, V., Lukose, R., Nagaraja, K., Pruyne, J., Richard, B., Rollins, S., and Xu, Z. (2002). Peer-to-Peer Computing. `http://www.hpl.hp.com/techreports/2002/`.

[Minar and Hedlund, 2001] Minar, N. and Hedlund, M. (2001). A Network of Peers: Peer-to-Peer Models Through the History of the Internet. In *Peer-to-Peer: Harnessing the Benefits of a Disruptive Technology*, pages 9–20.

[Mondal and Kitsuregawa, 2006] Mondal, A. and Kitsuregawa, M. (2006). Privacy, Security and Trust in P2P Environments: A Perspective. In *Proceedings 17th International Conference Database and Expert Systems Applications (DEXA 2006)*.

[μAmps Project, 2008] μAmps Project (2008). http://www-mtl.mit.edu/researchgroups/icsystems/uamps/.

[Nakajima et al., 2007] Nakajima, Y., Watanabe, K., and Nemati, A. G. (2007). Trustworthiness of Acquaintance Peers on Access Control Models. In *Proceedings 18th International Workshop on Database and Expert Systems Applications (DEXA)*.

[Naoumov and Ross, 2006] Naoumov, N. and Ross, K. (2006). Exploiting P2P Systems for DDoS Attacks. In *Proceedings of the First International Conference on Scalable Information Systems*.

[Napster, 2009] Napster (2009). `http://free.napster.com`.

[Newsome et al., 2004] Newsome, J., Shi, E., Song, D., and Perrig, A. (2004). The Sybil Attack in Sensor Networks: Analysis & Defenses. In *The 3rd International Symposium on Information Processing in Sensor Networks*, pages 259–268.

[Ohnishi et al., 2007] Ohnishi, K., Nagamatsu, S., Okamura, T., and Oie, Y. (2007). Autonomously Reconstructable Semi-Structured P2P Networks for File Sharing. In *Proceedings of the Third International Conference on Autonomic and Autonomous Systems (ICAS 2007)*.

[Oram, 2001] Oram, A. (2001). *Peer-to-Peer: Harnessing the Benefits of a Disruptive Technology*. Sebastopol.

[Osborne, 2004] Osborne, M. J. (2004). *An Introduction to Game Theory*. Oxford University Press.

[Paillier, 1999] Paillier, P. (1999). Public-Key Cryptosystems Based on Discrete Logarithm Residues. In *Proceedings Eurocrypt 1999*.

[Parno et al., 2005] Parno, B., Perrig, A., and Gligor, V. (2005). Distributed Detection of Node Replication Attacks in Sensor Networks. In *IEEE Symposium on Security and Privacy*, pages 49–63.

[Parvez et al., 2008] Parvez, K. N., Williamson, C., Mahanti, A., and Carlsson, N. (2008). Analysis of BitTorrent-Like Protocols for On-Demand Stored Media Streaming. In *Proceedings of ACM SIGNMETRICS 2008*, pages 301–312.

[Perrig et al., 2002] Perrig, A., Szewczyk, R., Tygar, J. D., Wen, V., and Culler, D. E. (2002). SPINS: Security Protocols for Sensor Networks. *Wireless Networks*, 8(5):521–534.

[Piatek et al., 2010] Piatek, M., Krishnamurthy, A., Venkataramani, A., Yang, R., Alex, D. Z., and Abstract, E. J. (2010). Contracts: Practical Contribution Incentives for P2P Live Streaming. In *Proceedings of NSDI 2010*.

[PlanetLab, 2006] PlanetLab (2006). `http://www.planet-lab.org`.

[PPLive, 2009] PPLive (2009). `http://www.pplive.com`.

[PPStream, 2009] PPStream (2009). `http://www.ppstream.com`.

[Qiu and Srikant, 2004] Qiu, D. and Srikant, R. (2004). Modeling and Performance Analysis of BitTorrent-Like Peer-to-Peer Networks. In *Proceedings of SIGCOMM*, pages 367–377.

[Qu et al., 2009] Qu, Z., Zhou, J., Harjula, E., and Ylianttila, M. (2009). Truncated Pyramid Peer-to-Peer Architecture with Vertical Tunneling Model. In *Proceedings of CCNC 2009*.

[Raghunathan et al., 2006] Raghunathan, V., Ganeriwal, S., and Srivastava, M. (2006). Emerging Techniques for Long Lived Wireless Sensor Networks. *IEEE Communications Magazine*, 44(4):108–114.

[Rajaraman, 2002] Rajaraman, R. (2002). Topology Control and Routing in Ad Hoc Networks: A Survey. In *ACM SIGACT News 2002*, pages 60–73.

[Ramaswamy and Liu, 2003] Ramaswamy, L. and Liu, L. (2003). Free Riding: A New Challenge to Peer-to-Peer File Sharing Systems. In *Proceedings of the 36th Hawaii International Conference on System Sciences*.

[Ranganathan et al., 2003] Ranganathan, K., Ripeanu, M., Sarin, A., and Foster, I. (2003). To Share or Not to Share: An Analysis of Incentives to Contribute in Collaborative File-Sharing Environments. In *Proceedings of the Workshop on Economics of Peer-to-Peer systems*.

[Ratnasamy et al., 2001] Ratnasamy, S., Francis, P., Handley, M., Karp, R., and Shenker, S. (2001). A Scalable Content-Addressable Network. In *Proceedings of ACM SIGCOMM 2001*.

[Ripenau, 2001] Ripenau, M. (2001). Peer-to-Peer Architecture Case Study: Gnutella Network. In *Proceedings of IEEE 1st International Conference on Peer-to-Peer Computing (P2P2001)*.

[Rivest et al., 1978] Rivest, R. L., Shamir, A., and Adleman, L. (1978). A Method for Obtaining Digital Signatures and Public-Key Cryptosystems. *Communications of the ACM*, 21:pp. 120–126.

[Rosenthal, 1964] Rosenthal, A. M. (1964). *Thirty-Eight Witnesses*. McGraw-Hill.

[Roussopoulos et al., 2004] Roussopoulos, M., Baker, M., Rosenthal, D., Giuli, T., Maniatis, P., and Mogul, J. (2004). 2 P2P or Not 2 P2P? In *Proceedings of the Third International Workshop on Peer-to-Peer Systems (IPTPS '04)*.

[Rowaihy et al., 2007] Rowaihy, H., Enck, W., McDaniel, P., and Porta, T. L. (2007). Limiting Sybil Attacks in Structured P2P Networks. In *Proceedings of the 2007 INFOCOM*, pages 2596–2600.

[Rowstron and Druschel, 2001a] Rowstron, A. and Druschel, P. (2001a). Pastry: Scalable, Decentralized Object Location and Routing for Large-Scale Peer-to-Peer Systems. In *Proceedings of the 18th IFIP/ACM International Conference on Distributed Systems Platforms (Middleware 2001)*.

[Rowstron and Druschel, 2001b] Rowstron, A. and Druschel, P. (2001b). Pastry: Scalable, Distributed Object Address and Routing for Large-Scale Peer-to-Peer Systems. In *Proceedings of the IFIP/ACM International Conference on Distributed Systems Platforms*.

[Saito, 2003] Saito, K. (2003). Peer-to-Peer Money: Free Currency over the Internet. In *Proceedings of the 2nd International Conference on Human.Society@Internet*.

[Saito et al., 2005] Saito, K., Morino, E., and Murai, J. (2005). Multiplication Over Time to Facilitate Peer-to-Peer Barter Relationship. In *Proceedings of the 16th International Workshop on Database and Expert Systems Applications*.

[Salem et al., 2006] Salem, N. B., Buttyan, L., Hubaux, J.-P., and Jakobsson, M. (2006). Node Cooperation in Hybrid Ad Hoc Networks. *IEEE Transactions on Mobile Computing*, 5(4):365–376.

[Sanghavi and Hajek, 2005] Sanghavi, S. and Hajek, B. (2005). A New Mechanism for the Free-Rider Problem. In *Proceedings of the SIGCOMM 2005 Workshop*.

[Schelling, 1971] Schelling, T. C. (1971). Dynamic Models of Segregation. *Journal of Mathematical Sociology*, (2).

[Schelling, 1978] Schelling, T. C. (1978). *Micromotives and Macrobehavior.* W. W. Norton & Company.

[Schmidt et al., 2007] Schmidt, S., Steele, R., and Dillon, T. (2007). DEco Arch: Trust and Reputation Aware Service Brokering Architecture in Digital Ecosystems. In *Proceedings Inaugural IEEE International Conference Digital Ecosystems and Technologies (DEST)*.

[Schoder and Fischbach, 2003] Schoder, D. and Fischbach, K. (2003). Peer-to-Peer Prospects. *Communications of the ACM*, (2):27–29.

[Schollmeier, 2002] Schollmeier, R. (2002). A Definition of Peer-to-Peer Networking for the Classification of Peer-to-Peer Architectures and Applications. In *Proceedings of the First International Conference on Peer-to-Peer Computing (P2P'01)*, pages 27–29.

[Seedorf, 2006] Seedorf, J. (2006). Security Challenges for Peer-to-Peer SIP. *IEEE Network*, pages 38–45.

[Sentinelli et al., 2007] Sentinelli, A., Marfia, G., Gerla, M., and Kleinrock, L. (2007). Will IPTV Ride the Peer-to-Peer Stream? *IEEE Communications Magazine*, pages 86–92.

[SETI@Home, 2009] SETI@Home (2009). `http://setiathome.berkeley.edu`.

[Shakkottai and Srikant, 2007] Shakkottai, S. and Srikant, R. (2007). Peer to Peer Networks for Defense Against Internet Worms. *IEEE Journal on Selected Areas in Communications*, (9):1745–1752.

[Singh and Haahr, 2006] Singh, A. and Haahr, M. (2006). Creating an Adaptive Network of Hubs Using Schelling's Model. *Communications of the ACM*, (3):69–73.

[Singh and Raghavendra, 1998] Singh, S. and Raghavendra, C. S. (1998). PAMAS—Power Aware Multi-Access Protocol with Signalling for Ad Hoc Networks. In *ACM SIGCOMM Computer Communication Review*, pages 5–26.

[Singh et al., 2003] Singh, S., Ramabhadran, S., Baboescu, F., and Snoeren, A. C. (2003). The Case for Service Provider Deployment of Super-Peers in Peer-to-Peer Networks. In *Proceedings of the Workshop on Economics of Peer-to-Peer Systems*.

[Singh et al., 1998] Singh, S., Woo, M., and Raghavendra, C. S. (1998). Power-Aware Routing in Mobile Ad Hoc Networks. In *Proceedings ACM MOBICOM 1998*, pages 181–190.

[Skype, 2009] Skype (2009). http://www.skype.com.

[Smart Dust Project, 2008] Smart Dust Project (2008). http://robotics.eecs.berkeley.edu/~pister/smartdust/.

[Smith et al., 2003] Smith, H., Clippinger, J., and Konsynski, B. (2003). Riding the Wave: Discovering the Value of P2P Technologies. *Communications of the Association for Information Systems*, pages 94–107.

[Song et al., 2005] Song, S., Hwang, K., Zhou, R., and Kwok, Y.-K. (2005). Trusted P2P Transactions with Fuzzy Reputation Aggregation. *IEEE Internet Computing*, pages 24–34.

[Staniford et al., 2002] Staniford, S., Paxon, V., and Weaver, N. (2002). How to Own the Internet in Your Spare Time. In *Proceedings of the 11th USENIX Security Symposium*.

[Steinmetz and Wehrle, 2005] Steinmetz, R. and Wehrle, K. (2005). *Peer-to-Peer Systems and Applications*. Springer.

[Stoica et al., 2001a] Stoica, I., Morris, R., Karger, D., Kaashoek, M. F., and Balakrishnan, H. (2001a). Chord: A Scalable Peer-to-Peer Lookup Service for Internet Applications. In *Proceedings of ACM SIGCOMM 2001*, pages 1–12.

[Stoica et al., 2001b] Stoica, I., Morris, R., Karger, D., Kaashoek, M. F., and Balakrishnan, H. (2001b). Chord: A Scalable Peer-to-Peer Lookup Service for Internet Applications. In *Proceedings of the ACM SIGCOMM 2001*, pages 149–160.

[Sun and Garcia-Molina, 2004] Sun, Q. and Garcia-Molina, H. (2004). SLIC: A Selfish Link-Based Incentive Mechanism for Unstructured Peer-to-Peer Networks. In *Proceedings of the 24th International Conference on Distributed Computing Systems*.

[Sung et al., 2008] Sung, W.-L., Hu, S.-Y., and Jiang, J.-R. (2008). Selection Strategies for Peer-to-Peer 3D Streaming. In *Proceedings of ACM NOSSDAV 2008*, pages 15–20.

[Suryanarayana et al., 2005] Suryanarayana, G., Erenkrantz, J. R., and Taylor, R. N. (2005). An Architectural Approach for Decentralized Trust Management. *IEEE Internet Computing*, pages 16–23.

[Tang et al., 2007] Tang, Y., Luo, J.-G., Zhang, Q., Zhang, M., and Yang, S. (2007). Deploying P2P Networks for Large-Scale Live Video-Streaming Service. *IEEE Communications Magazine*, pages 100–106.

[Tribler, 2009] Tribler (2009). http://www.tribler.org.

[Tuan, 2006] Tuan, T. A. (2006). A Game-Theoretic Analysis of Trust Management in P2P Systems. In *Proceedings CCE 2006*.

[Varian, 2003] Varian, H. R. (2003). The Social Cost of Sharing. In *Proceedings of the Workshop on Economics of Peer-to-Peer Systems*.

[Venot and Yan, 2007] Venot, S. and Yan, L. (2007). Peer-to-Peer Media Streaming Application Survey. In *Proceedings of the International Conference on Mobile Ubiquitous Computing, Systems, Services and Technologies*, pages 139–148.

[vivek Shrivastava and Banerjee, 2005] vivek Shrivastava and Banerjee, S. (2005). Natural Selection in Peer-to-Peer Streaming: From the Cathedral to the Bazaar. In *Proceedings of NOSSDAV'05*.

[Vu et al., 2010] Vu, L., Gupta, I., Nahrstedt, K., and Liang, J. (2010). Understanding Overlay Characteristics of a Large-Scale Peer-to-Peer IPTV System. *ACM Transactons on Multimedia Computing, Communications, and Applications*, (4).

[Wang and Bhargava, 2004] Wang, W. and Bhargava, B. (2004). Visualization of Wormholes in Sensor Networks. In *ACM Workshop on Wireless Security*, pages 51–60.

[Wang and Li, 2005] Wang, W. and Li, B. (2005). Market-Driven Bandwidth Allocation in Selfish Overlay Networks. In *Proceedings of the IEEE INFOCOM 2005*.

[Wang et al., 2005] Wang, Y., Reibman, A. R., and Lin, S. (2005). Multiple Description Coding for Video Delivery. *Proceedings of the IEEE*, pages 57–70.

[Watro et al., 2004] Watro, R., Kong, D., Cuti, S., Gardiner, C., Lynn, C., and Kruus, P. (2004). TinyPK: Securing Sensor Networks with Public Key Technology. In *The 2nd ACM Workshop on Security of Ad Hoc and Sensor Networks*, pages 59–64.

[Watts and Strogatz, 1998] Watts, D. and Strogatz, S. (1998). Collective Dynamics of Small-World Networks. *Nature*, (6684):409–410.

[Wei and Chen, 2008] Wei, T. and Chen, C. (2008). Study of PPStream Based on Measurement. In *Proceedings of the Second International Symposium on Intelligent Information Technology Application*, pages 900–905.

[Wikipedia, 2011] Wikipedia (2011). `http://www.wikipedia.org/`.

[WinMX World, 2009] WinMX World (2009). `http://winmxworld.com`.

[WINS Project, 2008] WINS Project (2008). http://www.janet.ucla.edu/wins/.

[Wolfson et al., 2004] Wolfson, O., Xu, B., and Sistla, A. P. (2004). An Economic Model for Resource Exchange in Mobile Peer to Peer Networks. In *Proceedings of the 16th International Conference on Scientific and Statistical Databased Management.*

[Wongrujira and Seneviratne, 2005] Wongrujira, K. and Seneviratne, A. (2005). Monetary Incentive with Reputation for Virtual Market-Place Based P2P. In *Proceedings of CoNEXT'05.*

[Wood and Stankovic, 2002] Wood, A. D. and Stankovic, J. A. (2002). Denial of Service in Sensor Networks. *IEEE Computer Magazine*, 35(10):54–62.

[Xie and Zhu, 2007] Xie, L. and Zhu, S. (2007). A Feasibility Study on Defending Against Ultra-Fast Topological Worms. In *Proceedings of the Seventh IEEE International Conference on Peer-to-Peer Computing*, pages 61–68.

[Xie et al., 2007] Xie, S., Li, B., Keung, G. Y., and Zhang, X. (2007). Cool-Streaming: Desing, Theory, and Practice. *IEEE Transactions on Multimedia*, (8):1661–1671.

[Xiong and Liu, 2004] Xiong, L. and Liu, L. (2004). PeerTrust: Supporting Reputation-Based Trust for Peer-to-Peer Electronic Communities. *IEEE Transactions on Knowledge and Data Engineering*, (7):843–857.

[Xu et al., 2002] Xu, D., Hefeeda, M., Hambrusch, S., and Bhargava, B. (2002). On Peer-to-Peer Media Streaming. In *Proceedings of the 22nd International Conference on Distributed Computing Systems.*

[Xu et al., 2007] Xu, Z., He, Y., and Deng, L. (2007). A Multilevel Reputation System for Peer-to-Peer Networks. In *Proceedings 6th International Conference Grid and Cooperative Computing (GCC).*

[Xue et al., 2004] Xue, G.-T., Li, M.-L., Deng, Q.-N., and You, J.-Y. (2004). Stable Group Model in Mobile Peer-to-Peer Media Streaming System. In *Proceedings of the IEEE International Conference on Mobile Ad Hoc and Sensor Systems*, pages 334–339.

[Yang and Garcia-Molina, 2003] Yang, B. and Garcia-Molina, H. (2003). PPay: Micropayments for Peer-to-Peer Systems. In *Proceedings of the 10th ACM Conference on Computer and Communication Security*, pages 300–310.

[Yang et al., 2005] Yang, M., Zhang, Z., Li, X., and Dai, Y. (2005). An Empirical Study of Free-Riding Behavior in the Maze P2P File-Sharing System. In *Proceedings of IPTPS'05.*

[Yang et al., 2008] Yang, S., Jin, H., Li, B., Liao, X., Yao, H., and Tu, X. (2008). The Content Pollution in Peer-to-Peer Live Streaming Systems:

Analysis and Implications. In *Proceedings of the 37th International Conference on Parallel Processing*, pages 652–659.

[Ye and Makedon, 2004] Ye, S. and Makedon, F. (2004). Collaboration-Aware Peer-to-Peer Media Streaming. In *Proceedings of MM'04*, pages 412–415.

[Yeung and Kwok, 2006a] Yeung, M. K. H. and Kwok, Y.-K. (2006a). A Game Theoretic Approach to Power Aware Wireless Data Access. *IEEE Transactions on Mobile Computing*, 5(8).

[Yeung and Kwok, 2006b] Yeung, M. K. H. and Kwok, Y.-K. (2006b). On Maximizing Revenue for Client-Server Based Wireless Data Access in the Presence of Peer-to-Peer Sharing. In *Proceedings of the 17th Annual IEEE International Symposium on Personal, Indoor, and Mobile Radio Communications (PIMRC'2006)*.

[Yeung and Kwok, 2008] Yeung, M. K. H. and Kwok, Y.-K. (2008). Energy Efficient Media Streaming in Wireless Hybrid Peer-to-Peer Systems. In *Proceedings of the 22nd IEEE International Parallel and Distributed Processing Symposium (IPDPS 2008)*.

[Yeung and Kwok, 2009] Yeung, M. K. H. and Kwok, Y.-K. (2009). On Game Theoretic Peer Selection for Resilient Peer-to-Peer Media Streaming. *IEEE Transactions on Parallel and Distributed Systems*, (10):1512–1525.

[Yeung, 2008] Yeung, R. (2008). *Information Theory and Network Coding*. Springer.

[Yiu et al., 2007] Yiu, W.-P. K., Jin, X., and Chan, S.-H. G. (2007). Challenges and Approaches in Large-Scale P2P Media Streaming. *IEEE Multimedia*, pages 50–59.

[YouTube, 2009] YouTube (2009). http://www.youtube.com.

[Yu and Singh, 2003] Yu, B. and Singh, M. P. (2003). Incentive Mechanisms for Peer-to-Peer Systems. In *Proceedings of the 2nd International Workshop on Agents and Peer-to-Peer Computing*.

[Yu et al., 2008] Yu, H., Kaminsky, M., Gibbons, P. B., and Flaxman, A. D. (2008). SybilGuard: Defending Against Sybil Attacks via Social Networks. *IEEE/ACM Transactions on Networking*, (3):576–589.

[Zhang et al., 2005a] Zhang, J., Liu, L., and Pu, C. (2005a). Constructing a Proximity-Aware Power Law Overlay Network. In *Proceedings of the IEEE GLOBECOM 2005*, pages 636–640.

[Zhang et al., 2007] Zhang, M., Zhang, Q., Sun, L., and Yang, S. (2007). Understanding the Power of Pull-Based Streaming Protocol: Can We Do Better? *IEEE Journal on Selected Areas in Communications*, (9):1678–1694.

[Zhang et al., 2005b] Zhang, X., Liu, J., Li, B., and Yum, T.-S. P. (2005b). CoolStreaming/DONet: A Data-Driven Overlay Network for Peer-to-Peer Live Media Streaming. In *Proceedings of INFOCOM 2005*, pages 2102–2111.

[Zhang and Fang, 2007] Zhang, Y. and Fang, Y. (2007). A Fine-Grained Reputation System for Reliable Service Selection in Peer-to-Peer Networks. *IEEE Transactions on Parallel and Distributed Systems*, (8):1134–1145.

[Zhao et al., 2004] Zhao, B. Y., Huang, L., Stribling, J., Rhea, S. C., Joseph, A. D., and Kubiatowicz, J. D. (2004). Tapestry: A Resilient Global-Scale Overlay for Service Deployment. *IEEE Journal on Selected Areas in Communications*, (1):41–53.

[Zhou and Hwang, 2007a] Zhou, R. and Hwang, K. (2007a). Gossip-Based Reputation Aggregation for Unstructured Peer-to-Peer Networks. In *Proceedings 22nd International Parallel and Distributed Symposium (IPDPS)*.

[Zhou and Hwang, 2007b] Zhou, R. and Hwang, K. (2007b). PowerTrust: A Robust and Scalable Reputation System for Trusted Peer-to-Peer Computing. *IEEE Transactions on Parallel and Distributed Systems*, (4):460–473.

Index

Printed and bound by CPI Group (UK) Ltd, Croydon, CR0 4YY

23/10/2024

01777671-0007